# Table of Contents

LINEAR AND NONLINEAR PROGRAMMING:

**An Introduction to Linear Methods in Mathematical Programming**

## ELLIS HORWOOD SERIES IN MATHEMATICS AND ITS APPLICATIONS

*Series Editor:* Professor G. M. BELL, Chelsea College, University of London

## Statistics and Operational Research Section

*Editor:* B. W. CONOLLY, Chelsea College, University of London


Beaumont, G. P. **Introductory Applied Probability**
Beaumont, G. P. **Basic Probability and Random Variables***
Conolly, B. W. **Techniques in Operational Research: Vol. 1, Queueing Systems**
Conolly, B. W. **Techniques in Operational Research: Vol. 2, Models, Search, Randomization**
French, S. **Sequencing and Scheduling: Mathematics of the Job Shop**
French, S. **Decision Theory**
Hartley R. **Linear Methods of Mathematical Programming**
Jones, A. J. **Game Theory**
Kemp, K. W. **Dice, Data and Decisions: Introductory Statistics**
Oliveira-Pinto, F. **Simulation Concepts in Mathematical Modelling**
Oliveira-Pinto, F. & Conolly, B. W. **Applicable Mathematics of Non-physical Phenomena**
Schendel, U. **Introduction to Numerical Methods for Parallel Computers**
Stoodley, K. D. C. **Applied and Computational Statistics A First Course**
Stoodley, K. D. C., Lewis, T. & Stainton, C. L. S. **Applied Statistical Techniques**
Thomas, L. C. **Games, Theory and Applications**
Whitehead, J. R. **The Design and Analysis of Sequential Clinical Trials**

******In preparation*

# LINEAR AND NONLINEAR PROGRAMMING:
## An Introduction to Linear Methods in Mathemathical Programming

ROGER HARTLEY, B.A., Ph.D.
Senior Lecturer in Decision Theory
University of Manchester

ELLIS HORWOOD LIMITED
Publishers · Chichester
Halsted Press: a division of
JOHN WILEY & SONS
New York · Chichester · Ontario · Brisbane

First published in 1985 by
**ELLIS HORWOOD LIMITED**
Market Cross House, Cooper Street, Chichester, West Sussex, PO19 1EB, England

*The publisher's colophon is reproduced from James Gillison's drawing of the ancient Market Cross, Chichester.*

**Distributors:**
*Australia, New Zealand, South-east Asia:*
Jacaranda-Wiley Ltd., Jacaranda Press,
JOHN WILEY & SONS INC.,
G.P.O. Box 859, Brisbane, Queensland 40001, Australia
*Canada:*
JOHN WILEY & SONS CANADA LIMITED
22 Worcester Road, Rexdale, Ontario, Canada.
*Europe, Africa:*
JOHN WILEY & SONS LIMITED
Baffins Lane, Chichester, West Sussex, England.
*North and South America and the rest of the world:*
Halsted Press: a division of
JOHN WILEY & SONS
605 Third Avenue, New York, N.Y. 10016, U.S.A.



**British Cataloguing in Publication Data**
Hartley, Roger
Linear and Nonlinear Programming:
An Introduction to Linear Methods in Mathematical Programming. –
(Ellis Horwood series in Mathematics and its Applications)
1. Linear Programming
I. Title 2. Series
519.7'2 T57.74

ISBN 0-85312-644-5 (Ellis Horwood Limited – Library Edn.)
ISBN 0-85312-848-0 (Ellis Horwood Limited – Student Edn.)
ISBN 0-470-20178-9 (Halsted Press – Library Edn.)
ISBN 0-470-20179-7 (Halsted Press – Student Edn.)

Typeset by Heather FitzGibbon, Fleet, nr. Aldershot, Hants.
Printed in Great Britain by The Camelot Press, Southampton

# Preface

This book deals with linear programming and a selection of other topics which can be handled by extending linear programming methods. It arose out of a course given to undergraduate and postgraduate students from a wide range of numerate disciplines. The minimal common mathematical background of these students imposed severe restrictions on the prior knowledge I could assume. In striving to avoid either excessive preliminary material or the trap of the 'cook-book', I have adopted an approach which is rigorous and complete but informal in presentation. The only mathematical prerequisites are an ability to handle equations and inequalities (knowledge of the theory of equations is not required) and familiarity with summation notation. However, the reader will be expected to follow arguments running, in some cases, over several chapters.

The text is also written to reflect some of the massive research effort that has been directed towards linear programming and related areas. In particular, aspects of linear programming computation, integer programming, network problems and multiple objective methods have undergone considerable development in the last decade. This work has influenced the choice of topics.

Informality of presentation means that, typically examples are solved *before* generalities are discussed. The student is strongly advised to study the worked examples carefully, even better, to try solving them himself – and then to attempt some or all of the exercises. These are not optional. The only way to really understand the material is by plenty of practice, both on routine and on more demanding exercises. Answers or hints are provided for all exercises.

The book emphasises theory and computational methods which are widely applied in all areas of industry and planning, and it is written with the idea of computer implementation in mind (the ambitious reader might even try writing his own code).

Chapters 1, 2, 3, 5 and 6 constitute the basic material on linear programming most of which is used in most of the subsequent chapters. Chapter 4 is more demanding, but not referred to again (except in exercises). Chapter 7 is used in Section 8.5 and Chapter 10 and the introductory section of Chapter 9 is referred to in Chapters 10 and 12. Apart from this, Chapters 8, 9, 10, 11 and 12 and even certain sections from within them, are independent and can be read or skipped as desired.

It gives me great pleasure to acknowledge the stimulation derived from Doug White, Lyn Thomas, and Simon French of the Department of Decision Theory at the University of Manchester and to thank Regina Benveniste for discussions on quadratic programming. My students have offered many helpful and valuable suggestions over the years. All these have shaped my thoughts on mathematical programming and how it should be presented. Some of the exercises use parts (sometimes modified) of exam questions set for students at the University of Manchester and permission to use them is gratefully acknowledge. Finally, a *sine qua non*, the excellent typing of Jill Weatherall is much appreciated.

*April 1984* Roger Hartley

# Selective Index of Notation

This index does not include 'local notation' used temporarily within a short section of the text, or the different characters used to represent 'variables' in particular contexts.

| Symbol | Explanation | First introduced on page |
|---|---|---|
| $s_i$ | slack variable | 14 |
| $x_{B_i}$ | $i$th basic variable | 22 |
| $x_{N_j}$ | $j$th non-basic variable | 22 |
| $\alpha_{ij}$ | coefficient of $x_{N_j}$ in $i$th equation | 22 |
| $z$ | objective function | 36 |
| $a_i$ | artificial variable | 41 |
| $c_{B_i}(c_{N_j})$ | objective function coefficient of $x_{B_i}(x_{N_j})$ | 46 |
| $\beta_{ij}$ | coefficient of $X_j$ in $i$th equation | 55 |
| $\eta_i$ | $i$th element in $\eta$-list | 56 |
| $\pi_i$ | output of backwards transformation | 57 |
| $\delta_k$ | reciprocal of column weighting factor | 62 |
| $w$ | minus dual objective function | 68 |
| $t_j$ | dual slack variable | 70 |
| $b_{B_i}(b_{N_j})$ | RHS of constraint in which $x_{B_i}(x_{N_j})$ is slack | 88 |
| $U_J$ | upper bound on $x_J$ | 100 |
| $\sigma_J$ | slack variable associated with $x_J$ | 100 |
| $L_J$ | lower bound on $x_J$ | 104 |
| $P(j)$ | parent (vertex) of $j$ | 107 |
| $w_k$ | $k$th objective function weight | 129 |
| $\gamma_{kj}$ | canonical coefficient of $x_{N_j}$ in $k$th objective | 138 |
| $d_k(e_k)$ | deviation variables | 148 |
| $v_{\mathrm{I}}(v_{\mathrm{II}})$ | optimal value of a game to player I(II) | 155 |
| $z_k$ | optimal objective value of $k$th subproblem | 167 |
| $f_{i0}$ | fractional part of $\alpha_{i0}$ | 168 |
| $D_k(.)$ | down-penalty | 168 |

| | | |
|---|---|---|
| $U_k(.)$ | up-penalty | 169 |
| $\phi$ | PWL approximation | 176 |
| $\lambda_i(a_k)$ | variables (breakpoints) in SOS | 177 |
| $\lambda_i(\mu_j)$ | Kuhn–Tucker multiplier associated with $s_i(x_j)$ | 186 |

CHAPTER 1

# Introduction

---

## 1.1 AN EXAMPLE – RECYCLING WASTE PAPER

The primary impetus for the study of **Linear Programming Problems** (LPs) is the wide range of practical problems which can be, or have been, modelled as LPs. LP modelling constitutes a study in itself (see Further Reading) and, as our intention is to concentrate on solution procedures, we will limit the discussion to a single example. This is a simplified version of a model designed to explore some of the possible benefits from recycling waste paper.

Let us assume that the paper-making industry in a certain country makes, say, twelve different types of paper and that current annual production of type $i$ is $p_i$ tons ($i = 1, \ldots, 12$). Some of the paper produced is lost permanently from the system – exported, burnt, stored in libraries in the form of books, etc. – and the remaining waste paper, $v_i$ tons of type $i$, has to be disposed of. One possible means of mitigating the disposal problem is to recycle some of this waste into secondary pulp, which can then be used as part of the input to the production process, the other ingredient being virgin pulp. The technology of paper-making imposes a lower limit, $\alpha_i$ for paper of type $i$, on the proportion of input which must be virgin pulp. The costs of collecting and processing waste paper into secondary pulp should also be taken into account, but such costs can be very difficult to measure and so we will seek to determine how much paper should be recycled in order to minimise the total amount of virgin pulp used, when a proportion $\lambda$ of the total waste paper is available for recycling. Fulfilling this objective will also minimise the residual amount of waste paper to be disposed of.

Let us define $y_i(z_i)$ to be the amount of virgin (secondary) pulp used as input to the production of paper of type $i$ ($i = 1, \ldots, 12$) and $w_{ij}$ to be the amount of waste paper of type $i$ used in the production of paper of type $j$ in a year (measured in tons). If a 5 per cent weight loss is involved in the production process,

$$0.95y_i + 0.95z_i = p_i, \quad i = 1, \ldots, 12. \tag{1.1}$$

The minimum virgin pulp requirement can be restated as

$$0.95y_i \geqslant \alpha_i p_i, \qquad i = 1, \ldots, 12. \tag{1.2}$$

The production process dictates that only certain waste papers can be used in the production of other papers. So let us put $a_{ij} = 1$, if waste paper of type $i$ can be used in secondary pulp for the production of paper of type $j$, and $a_{ij} = 0$, otherwise. Then we must have

$$z_j = \sum_{i=1}^{12} a_{ij} w_{ij}, \qquad j = 1, \ldots, 12, \tag{1.3}$$

and, since $\lambda v_i$ tons of paper of type $i$ are available for recycling,

$$\sum_{j=1}^{12} a_{ij} w_{ij} \leqslant \lambda v_i, \qquad i = 1, \ldots, 12, \tag{1.4}$$

for the left-hand side of (1.4) is the total amount of waste paper of type $i$ consumed. When $a_{ij} = 0$, $w_{ij}$ is absent from (1.3)/(1.4).

Our objective is to minimise total virgin pulp used, i.e.

$$\sum_{i=1}^{12} y_i \tag{1.5}$$

whilst also satisfying $y_i$, $z_i$, $w_{ij} \geqslant 0$ for $i, j = 1, \ldots, 12$ and (1.1. . .(1.4). This example exhibits the typical features of an LP: a linear **objective function** (1.5), to be maximised or minimised subject to linear restrictions, or **constraints** (1.1) –(1.4) on the non-negative variables. Any non-negative solution of the constraints is called a **feasible solution.** Any feasible solution maximising or minimising the objective function is called **optimal**.

Some of the constraints are inequalities, such as (1.2) and (1.4); the remainder are equalities. Inequality constraints can always be converted to equalities by adding or subtracting a non-negative variable. Thus (1.2) can be rewritten

$$0.95y_i - s_i = \alpha_i p_i, \qquad i = 1, \ldots, 12,$$

where $s_i \geqslant 0$, and (1.4) can be rewritten

$$\sum_{j=1}^{12} a_{ij} w_{ij} + t_i = \lambda v_i, \qquad i = 1, \ldots, 12,$$

where $t_i \geqslant 0$. The variables introduced into the constraints are called **slack** variables and often have a natural interpretation in the model. For example, $t_i$ above is the amount of waste paper of type $i$, potentially available for recycling, that is not actually used. Another useful trick for standardising problems is based on the observation that minimising a function gives the same optimal solution(s) as maximising its negative, so that, in our waste paper problem, we could have chosen to maximise

$$-\sum_{i=1}^{12} y_i.$$

We can now write the LP as

$$\text{P1:} \qquad \text{maximise} \sum_{i=1}^{12} -y_i$$

$$\begin{aligned}
\text{subject to} \quad & 0.95y_i + 0.95z_i = p_i, & i = 1, \ldots, (12) \\
& 0.95y_i - s_i = \alpha_i p_i, & i = 1, \ldots, (12) \\
& \sum_{i=1}^{12} a_{ij} w_{ij} - z_j = 0, & j = 1, \ldots, (12) \\
& \sum_{j=1}^{12} a_{ij} w_{ij} + t_i = \lambda v_i, & i = 1, \ldots, (12) \\
& y_i, z_i, s_i, w_{ij}, z_j, t_i \geqslant 0, & i, j = 1, \ldots, (12).
\end{aligned}$$

Any LP, such as P1, which is written so that its objective function must be maximised and with only equality constraints (in non-negative variables) is said to be in **standard equality form** and this form will prove particularly valuable when we come to the development of computational procedures.

Occasionally, problems arise in which not all the variables are required to be non-negative. Such unrestricted variables are called **free** variables and it is straightforward to modify computational procedures to accommodate them (see Exercise 2 of Chapter 3). Alternatively, if $x_j$ is a free variable we can write $x_j = y_j - z_j$ (i.e. substitute $y_j - z_j$ throughout the problem for $x_j$) where $y_j$, $z_j \geqslant 0$. In this way, at the expense of introducing extra variables, we can convert the problem to standard equality form. In some problems there may be both free variables and equality constraints and one can adopt the strategy of using the equality constraint to express a free variable in terms of other variables and thereby eliminating it from the other constraints and the objective function (see Exercise 3). This has the advantage that the numbers of contraints and variables are both reduced by one.

## 1.2 A GRAPHICAL METHOD

For the rest of this chapter we will concentrate on LPs with two variables and describe a graphical procedure for solving them. Since such problems are unlikely to arise in realistic models, the procedure is offered, not as a serious competitor to more sophisticated methods but, rather, to illuminate some essential features of linear optimisation. This geometrical approach can be developed into a systematic methodology of linear programming, but the mathematical level involved would exceed the limits set for this book. In any case, such a development is probably better employed in the elucidation of non-linear (especially convex) programming. Instead, we will adopt an algebraic and computational approach, but geometric ideas will sometimes be used to provide an alternative viewpoint on important concepts.

To start, we shall examine the problem P2.

P2: maximise $-2x_1 + 3x_2$ $(=z)$

subject to

$$2x_1 - x_2 \leqslant 4 \quad \text{(I)}$$

$$x_1 - 2x_2 \geqslant -2 \quad \text{(II)}$$

$$2x_1 + x_2 \geqslant 2 \quad \text{(III)}$$

$$x_1, x_2 \geqslant 0$$

In which we have labelled the constraints for future use. We will ignore the objective function for the moment. The constraints are represented in Fig. 1.1

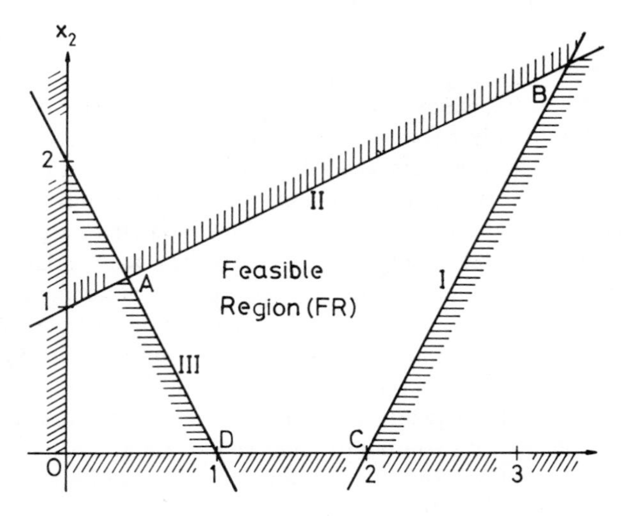


Figure 1.1 Feasible region of P2.

The equality corresponding to constraint I: $2x_1 - x_2 = 4$, gives the line marked I in the figure. The points $(x_1, x_2)$ satisfying the first constraint lie on, or to one side of, this line. To decide which side, we need only substitute a point obviously on one side of the line and see if the constraint is satisfied. The origin (0, 0) is usually the obvious choice and in the present case shows that points to the left of the line I satisfy the constraint. This is indicated in the figure by hatching the side of the line *not* satisfying the constraint, that is those points $(x_1, x_2)$ for which $2x_1 - x_2 > 4$. The same has been done for the second and third constraints. In addition, the $x_1$-axis has been hatched below to indicate the constraint $x_2 \geqslant 0$ and the left-hand side of the $x_2$-axis hatched to indicate $x_1 \geqslant 0$. The set of feasible points or solutions, called the **feasible region**, is the quadrilateral ABCD (including its interior).

The feasible region is redrawn in Fig. 1.2, with the hatching omitted. Also included in this figure is a series of lines on which we have set the objective function equal to various constants. Since the coefficients in the objective

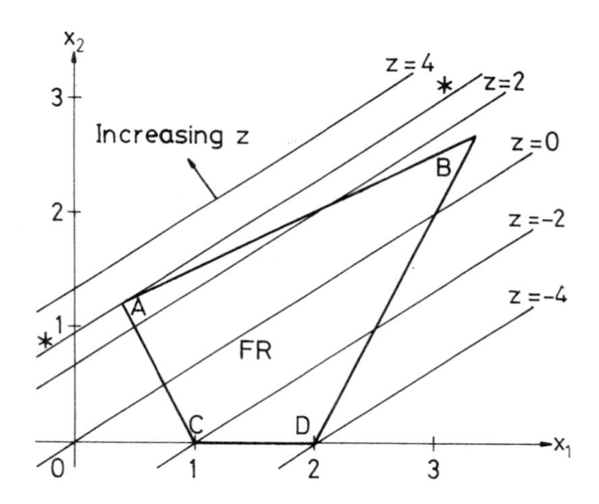


Figure 1.2 Graphical solution of P2.

function do not change, the lines are all parallel. The line $z = 0$ includes points which are feasible and the same is true of $z = 2$, improving the objective function. However, we cannot improve the objective function as far as $z = 4$, since this line includes no feasible points. The best we can achieve is the line indicated with asterisks. This clearly includes the feasible point A, but no feasible points can be found above this line. The optimal solution must be A, which lies on the lines II and III, and must, therefore, satisfy

$$x_1 - 2x_2 = -2$$

$$2x_1 + x_2 = 2,$$

which has the solution $(x_1, x_2) = (\frac{2}{5}, 1\frac{1}{5})$. The maximal objective function value is $z = -2 \times \frac{2}{5} + 3 \times \frac{6}{5} = 2\frac{4}{5}$.

As another example, we shall examine

$$\text{P3:} \quad \text{minimise} \quad 2x_1 + 6x_2$$

$$\text{subject to} \quad x_1 + 3x_2 \geqslant 3 \qquad \text{(I)}$$

$$2x_1 - x_2 \geqslant 2 \qquad \text{(II)}$$

$$x_1, x_2 \geqslant 0$$

which is solved in Fig. 1.3

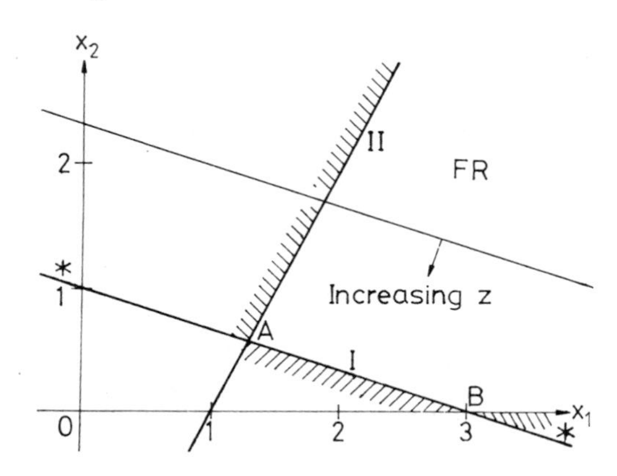


Figure 1.3 Graphical solution of P3.

In this example the objective function line is parallel to the edge AB of the feasible region. Consequently, both A = $(1\frac{2}{7}, \frac{4}{7})$ and B = $(3, 0)$ are optimal, as are all intermediate points on the line joining A and B. When more than one feasible solution is optimal, we will say that there are **alternative optimal solutions**. Another feature distinguishing Fig. 1.3 from Fig. 1.2 is that the feasible region is **unbounded.** Intuitively, this means that it is unlimited in extent but, more formally, we can say that the feasible region is **bounded** if upper limits $U_j$ on feasible $x_j$ can be found, that is $x_j \leqslant U_j$, for all $j$ and all feasible $x_j$. In the case of P2, Fig. 1.1 shows that $U_1 = 4$ and $U_2 = 3$ would do. If no such upper limit can be found (as in P3), the feasible region is unbounded. As we have seen in P3, an unbounded feasible region does not preclude an optimal solution and it may occur in perfectly well-formulated problems. However, if the objective in P3 had been to *maximise* $2x_1 + 6x_2$, it is clear that no optimal solution could have been found, for the objective function could have been increased without limit. In such a situation, we describe the LP as **unbounded** and, if such a problem arose in practice, we would suspect an error in the model. Notice the distinction between applying the epithet 'unbounded' to the feasible region and to the LP. An unbounded LP must have an unbounded feasible region but the converse is not necessarily true.

## 1.3 VERTICES

The feasible regions we have seen in problems P2 and P3 have exhibited a finite number of corners or **vertices**, for example A, B, C and D in Fig. 1.1 and A and B in Fig. 1.3. We are assuming that the reader has an intuitive notion of vertex, although a formal definition could be constructed. Problems P2 and P3 have also verified the assertion that *every LP with an optimal solution has a vertex which* *is optimal* and indeed, when vertices are rigorously defined, this assertion can be proved, and not just for problems with two variables. We shall not prove the assertion but we will discover how to recognise vertices algebraically and will eventually see that such an algebraic equivalent must be optimal, if an optimal solution exists. This is an important result for it reduces the problem of finding the best solution in the feasible region (an infinite set) to that of finding the best vertex (a finite set). Indeed, the simplex method, which forms the basis of all our subsequent development, can be viewed as a systematic procedure for searching the vertices.

Notice that an optimal solution need not be a vertex, in general, for the existence of alternative optima implies that there are optimal edge solutions; for example, points on the line between A and B in Fig. 1.3 are optimal, but not vertices. A *unique* optimal solution must be a vertex (why?) but this result is not particularly useful, since it is not usually possible to preclude alternative optima, prior to solving the problem.

Furthermore, it is not necessary for every LP to possess an optimal vertex since, if the problem is unbounded, there is no optimal solution at all. Another reason for the absence of an optimal solution is illustrated by the LP:

$$
\begin{array}{llll}
\text{P4:} & \text{maximise} & 2x_1 - 5x_2 & \\
& \text{subject to} & 2x_1 - \phantom{2}x_2 \leqslant 1 & \text{(I)} \\
& & -x_1 + 2x_2 \leqslant -1 & \text{(II)} \\
& & -2x_1 - \phantom{2}x_2 \leqslant -1 & \text{(III)} \\
& & x_1, x_2 \geqslant 0. &
\end{array}
$$

If we multiply (I) and (II) by 2 and add the resulting inequalities to (III) we obtain $x_2 \leqslant -1$ and any feasible solution must satisfy this. However, $x_2 \leqslant -1$ contradicts the requirement $x_2 \geqslant 0$. In other words there are no feasible solutions to P4 and, *a fortiori*, no vertices. In this case we will say that the LP is **infeasible.**

## EXERCISES

1. A chemical plant can use either of two processes but not both simultaneously. Process I uses 2 tons of raw material $A$ and 7 tons of raw material $B$ per day. It produces 3 tons of chemical $X$ and 6 tons of chemical $Y$, per day. Process II uses 5 tons of $A$ and 2 tons of $B$ and produces 2 tons of $X$ and 5 tons of $Y$, per day. Each month, up to 110 tons of $A$ and 140 tons of $B$ are available at a cost of £5,000 and £7,000 per ton, respectively. $X$ and $Y$ sell at £18,000 and £3,000 per ton respectively. Taking a month to be 30 days and assuming that all chemicals produced can be sold, formulate, as an LP in two variables, the problem of deciding how many days in each month each of the two processes should be used. Solve the problem graphically.

2. Rewrite the following LP in standard equality form, with three constraints. (Write free variables as the difference of signed variables.)

$$\begin{aligned}
\text{minimise} \quad & 5x_1 - x_2 - x_3 \\
\text{subject to} \quad & x_1 + x_2 + 7x_3 \geqslant 19 \\
& 5x_1 - x_2 + 12x_3 \leqslant 10 \\
& 7x_1 + x_2 - 3x_3 = 5 \\
& x_2 \geqslant 0,\ x_1,\ x_3 \ \text{free.}
\end{aligned}$$

3. Use elimination to reduce the following LP to one with two variables and three constraints and hence show, graphically, that it has alternative optimal solutions:

$$\begin{aligned}
\text{minimise} \quad & 2x_1 + 4x_2 + 3x_3 + 7x_4 \\
\text{subject to} \quad & x_1 + x_2 + x_3 + x_4 = 1 \\
& 4x_1 + 2x_2 + 5x_3 + x_4 = 2 \\
& -7x_1 - 7x_2 - 9x_3 - 9x_4 \leqslant 7 \\
& -x_1 + x_2 - x_3 + x_4 \geqslant 0 \\
& x_1 + 5x_2 + x_3 - 8x_4 \geqslant -3 \\
& \text{All variables are free.}
\end{aligned}$$

Write down three different optimal solutions.

4. Solve the following LP, (i) for $\theta = 3$ and (ii) for all values of $\theta$,

$$\begin{aligned}
\text{maximise} \quad & x_1 - \theta x_2 \\
\text{subject to} \quad & x_1 - 2x_2 \leqslant 3 \\
& -3x_1 + x_2 \leqslant 2 \\
& 2x_1 - x_2 \leqslant 10 \\
& x_1, x_2 \geqslant 0.
\end{aligned}$$

5. Is it true that an LP with alternative optimal solutions must always have at least two optimal vertices? Justify your answer graphically.

CHAPTER 2

# Systems of Equations

## 2.1 CANONICAL FORM

The constraints in the standard equality form of the linear programming problem define a system of linear equations. The reader will no doubt have some experience of solving linear equations when the numbers of variables and equations are the same. However, in linear programming, the number of variables typically exceeds the number of equations and in this chapter we shall examine such systems. One obvious procedure would be to select as many variables as there are equations and 'solve' for these variables in terms of the remaining variables (if possible). The result could be written in a form similar to the following example,

$$\begin{aligned}
E_1 &: \quad x_3 \qquad\qquad\qquad -2x_2 + x_5 = 1 \\
E_2 &: \qquad\quad x_1 \qquad\qquad +x_2 \qquad\quad = -7 \\
E_3 &: \qquad\qquad\quad x_4 \quad +3x_2 - 4x_5 = 9.
\end{aligned}$$

The defining feature of such a system is that each equation has an associated variable, called **basic**, which occurs in that equation alone and has coefficient 1. The collection of basic variables, $\{x_1, x_3, x_4\}$ in the example, will be called the **basis**. The remaining variables are **non-basic** and the system is said to be in **canonical form** with respect to the given basis.

Rather than deal with the equations of the example explicitly, we can represent them in a tabular form, called a **tableau** as in T1 below:

| | $x_1$ | $x_2$ | $x_3$ | $x_4$ | $x_5$ | T1 | |
|---|---|---|---|---|---|---|---|
| $x_3$ | 0 | −2 | 1 | 0 | 1 | 1 | Row 1 |
| $x_1$ | 1 | 1 | 0 | 0 | 0 | −7 | Row 2 |
| $x_4$ | 0 | 3 | 0 | 1 | −4 | 9 | Row 3 |

The three rows correspond to the three equations. The entries in the right-hand, or **resource**, column are the Right-Hand Sides (RHSs) of the equations and the entries in the remaining, or **inner**, columns are the coefficients of the variable at the head of the column. We have also labelled each row, in the left-hand column, with the basic variable in the equation corresponding to that row.

From the definition of canonical form, we know that the entries in a column headed by a basic variable must contain a 1 (in the row in which it is basic) and 0s elsewhere. The entries in these columns are redundant and we can eliminate them to produce the **reduced tableau** T1$^*$.

| | $x_2$ | $x_5$ | T1$^*$ |
|---|---|---|---|
| $x_3$ | −2 | 1 | 1 |
| $x_1$ | 1 | 0 | −7 |
| $x_4$ | 3 | −4 | 9 |

In general, we shall write $x_{B_i}$ for the $i$th basic variable and $x_{N_j}$ for the $j$th non-basic variable. Thus, in the example,

$$x_{B_1} = x_3; \quad x_{B_2} = x_1; \quad x_{B_3} = x_4; \quad x_{N_1} = x_2; \quad x_{N_2} = x_5 .$$

The $i$th equation can then be written

$$x_{B_i} + \sum_j \alpha_{ij} x_{N_j} = \alpha_{io} , \tag{2.1}$$

where the sum is taken over all the non-basic variables. This means that, in our example, we would have as typical values:

$$\alpha_{11} = -2; \quad \alpha_{22} = 0; \quad \alpha_{20} = -7; \quad \alpha_{30} = 9.$$

The corresponding tabular form is most neatly displayed by rearranging the columns so that the basic variables, in order, come first and are followed by the non-basic variables. This tableau and its reduced version, T and T$^*$, are given below. We have exhibited a typical basic variable $x_{B_i}$ and non-basic variable $x_{N_j}$.

| | $x_{B_1}$ | .... | $x_{B_i}$ | .... | $x_{N_1}$ | .... | $x_{N_j}$ | .... | T |
|---|---|---|---|---|---|---|---|---|---|
| $x_{B_1}$ | 1 | .... | 0 | .... | $\alpha_{11}$ | .... | $\alpha_{1j}$ | .... | $\alpha_{10}$ |
| ⋮ | ⋮ | | ⋮ | | ⋮ | | ⋮ | | ⋮ |
| $x_{B_i}$ | 0 | .... | 1 | ..... | $\alpha_{i1}$ | .... | $\alpha_{ij}$ | .... | $\alpha_{io}$ |
| ⋮ | ⋮ | | ⋮ | | ⋮ | | ⋮ | | ⋮ |

It is worth noting that $j = 0$ (as in $\alpha_{io}$) refers to the resource column. Furthermore, the $x_{B_i}$, $x_{N_j}$ and $\alpha_{ij}$ all pertain to the canonical form or tableau corresponding to a particular basis. If another basis is chosen, the $x_{B_i}$, $x_{N_j}$ and $\alpha_{ij}$ will change.

| | $x_{N_1}$ .... $x_{N_j}$ .... | $T^*$ |
|---|---|---|
| $x_{B_1}$ | $\alpha_{11}$ .... $\alpha_{ij}$ .... | $\alpha_{10}$ |
| ⋮ | ⋮ ⋮ | ⋮ |
| $x_{B_i}$ | $\alpha_{i1}$ .... $\alpha_{ij}$ .... | $\alpha_{io}$ |
| ⋮ | ⋮ ⋮ | ⋮ |

It is important to observe that the basis uniquely determines the canonical equations and therefore the tableau, for, if we had, in addition to (2.1),

$$x_{B_i} + \sum_j \overset{*}{\alpha}_{ij} x_{N_j} = \overset{*}{\alpha}_{io} \qquad \text{for } i \geqslant 1, \tag{2.2}$$

then (2.1) and (2.2) have a solution for any $x_{N_j}(j \geqslant 1)$ and therefore so does their difference:

$$\sum_j (\alpha_{ij} - \overset{*}{\alpha}_{ij}) x_{N_j} = \alpha_{io} - \overset{*}{\alpha}_{io} \quad \text{for } i \geqslant 1.$$

Substituting $x_{N_j} = 0$, for all $j$ gives $\alpha_{io} = \overset{*}{\alpha}_{io}$ for $i \geqslant 1$ and substituting $x_{N_k} = 1$ and $x_{N_j} = 0$ for all $j \neq k$ gives

$$\alpha_{ik} - \overset{*}{\alpha}_{ik} = \alpha_{io} - \overset{*}{\alpha}_{io} = 0 \text{ so } \alpha_{ik} = \overset{*}{\alpha}_{ik} \text{ for all } i, k \geqslant 1.$$

This means (2.2) is identical with (2.1).

## 2.2 CHANGING BASES – PIVOTING

Let us suppose we now wish to obtain the canonical form of our example, but with $\{x_1, x_2, x_3\}$ as basis. Since $x_2$ is not basic in the original, canonical form, it is necessary to associate it with some equation. Because $x_1$ and $x_3$ are remaining basic and appear in $E_1$ and $E_2$, the natural choice is to make $x_2$ basic in $E_3$. It must have a coefficient 1 and this is achieved by dividing $E_3$ by 3

$$E_3' = E_3/3: \qquad x_2 + \tfrac{1}{3}x_4 - \tfrac{4}{3}x_5 = 3.$$

However, $x_2$ still occurs in $E_1$ and $E_2$ and, to complete the canonical form, we must eliminate it from these equations. This can be achieved by multiplying $E_3'$ by the coefficient of $x_2$ in $E_i$ and subtracting the result from $E_3$. We find

$$E_1' = E_1 - (-2) \times E_3': \; x_3 \qquad + \tfrac{2}{3}x_4 - \tfrac{5}{3}x_5 = 7$$

$$E_2' = E_2 - E_3': \qquad x_1 \quad - \tfrac{1}{3}x_4 + \tfrac{4}{3}x_5 = -10.$$

The same transformation can be effected in tableau format by remembering that each row displays the coefficients and RHS of a particular equation and, therefore, rows can be multiplied by non-zero constants and added to and subtracted from each other. In this way we can summarise the calculations above in the tableau T2 which is obtained from T1 by the operations described. The reader should note that new row 3 has to be calculated first.

| | $x_1$ | $x_2$ | $x_3$ | $x_4$ | $x_5$ | T2 | |
|---|---|---|---|---|---|---|---|
| $x_3$ | 0 | 0 | 1 | $\frac{2}{3}$ | $-\frac{5}{3}$ | 7 | new row 1 = row 1 $-(-2)\times$ new row 3 |
| $x_1$ | 1 | 0 | 0 | $-\frac{1}{3}$ | $\frac{4}{3}$ | $-10$ | new row 2 = row 2 $-$ new row 3 |
| $x_2$ | 0 | 1 | 0 | $\frac{1}{3}$ | $-\frac{4}{3}$ | 3 | new row 3 = (row 3)/3 |

The transformation just discussed involved interchanging one basic with one non-basic variable ($x_2$ and $x_4$ in the example) and such interchanges will prove of great importance. We will, therefore, develop a set of rules for carrying out this transformation, so that the operations could be carried out on a computer. The rules will also turn hand computation into a simple mechanical exercise.

Let us assume that, in a general tableau, we have arranged the columns with basic variables first and that we wish to make $x_{B_r}$ non-basic and $x_{N_k}$ basic. The initial tableau then looks like IT.

| | $x_{B_1}$ | ... | $x_{B_r}$ | ... | $x_{B_i}$ | ... | $x_{N_1}$ | ... | $x_{N_k}$ | ... | $x_{N_j}$ | ... | IT |
|---|---|---|---|---|---|---|---|---|---|---|---|---|---|
| $x_{B_1}$ | 1 | ... | 0 | ... | 0 | ... | $\alpha_{11}$ | ... | $\alpha_{1k}$ | ... | $\alpha_{1j}$ | ... | $\alpha_{10}$ |
| ⋮ | ⋮ | | ⋮ | | ⋮ | | ⋮ | | ⋮ | | ⋮ | | ⋮ |
| $x_{B_r}$ | 0 | ... | 1 | ... | 0 | ... | $\alpha_{r1}$ | ... | $\alpha_{rk}$ | ... | $\alpha_{rj}$ | ... | $\alpha_{ro}$ |
| ⋮ | ⋮ | | ⋮ | | ⋮ | | ⋮ | | ⋮ | | ⋮ | | ⋮ |
| $x_{B_i}$ | 0 | ... | 0 | ... | 1 | ... | $\alpha_{i1}$ | ... | $\alpha_{ik}$ | ... | $\alpha_{ij}$ | ... | $\alpha_{io}$ |
| ⋮ | ⋮ | | ⋮ | | ⋮ | | ⋮ | | ⋮ | | ⋮ | | ⋮ |

To make $x_{N_k}$ basic we must divide the $x_{B_r}$-row by $\alpha_{rk}$ (and, therefore, require $\alpha_{rk} \neq 0$). Then, to derive the new $x_{B_i}$-row we must multiply the new $x_{B_r}$-row by $\alpha_{ik}$ (the coefficient of $x_{N_k}$) and subtract the result from the old $x_{B_i}$-row. The result of these calculations is shown below, in the new tableau NT.

We can now swamp the columns headed by $x_{B_r}$ and $x_{N_k}$ so that the basic variables again precede the non-basic variables and, by comparing the two tableaux, write down a set of rules to perform the transformation. We must first

| | $x_{B_1}$ | … | $x_{B_r}$ | … | $x_{B_i}$ | … | $x_{N_1}$ | … | $x_{N_k}$ | … | $x_{N_j}$ | … | NT |
|---|---|---|---|---|---|---|---|---|---|---|---|---|---|
| $x_{B_1}$ | 1 | … | $-\alpha_{1k}/\alpha_{rk}$ | … | 0 | … | $\alpha_{11}-\dfrac{\alpha_{r1}\,\alpha_{1k}}{\alpha_{rk}}$ | … | 0 | | $\alpha_{1j}-\dfrac{\alpha_{rj}\alpha_{1k}}{\alpha_{rk}}$ | … | $\alpha_{10}-\dfrac{\alpha_{r0}\,\alpha_{1k}}{\alpha_{rk}}$ |
| ⋮ | ⋮ | | ⋮ | | ⋮ | | ⋮ | | ⋮ | | ⋮ | | ⋮ |
| $x_{N_k}$ | 0 | … | $1/\alpha_{rk}$ | … | 0 | … | $\alpha_{r1}/\alpha_{rk}$ | … | 1 | … | $\alpha_{rj}/\alpha_{rk}$ | … | $\alpha_{r0}/\alpha_{rk}$ |
| ⋮ | ⋮ | | ⋮ | | ⋮ | | ⋮ | | ⋮ | | ⋮ | | ⋮ |
| $x_{B_i}$ | 0 | … | $-\alpha_{ik}/\alpha_{rk}$ | … | 1 | … | $\alpha_{i1}-\dfrac{\alpha_{r1}\,\alpha_{ik}}{\alpha_{rk}}$ | … | 0 | … | $\alpha_{ij}-\dfrac{\alpha_{rj}\alpha_{ik}}{\alpha_{rk}}$ | … | $\alpha_{i0}-\dfrac{\alpha_{r0}\,\alpha_{ik}}{\alpha_{rk}}$ |
| ⋮ | ⋮ | | ⋮ | | ⋮ | | ⋮ | | ⋮ | | ⋮ | | ⋮ |

re-label the variables by using the rules (primed variables refer to the transformed tableau):

$$x'_{B_r} = x_{N_k}; \qquad x'_{B_i} = x_{B_i} \quad \text{for } i \neq r$$

$$x'_{N_k} = x_{B_r}; \qquad x'_{N_j} = x_{N_j} \quad \text{for } j \neq k,$$

or, expressed more simply in verbal terms, we interchange the new basic and non-basic variables. The new entries are given by

$$\alpha'_{rk} = 1/\alpha_{rk}; \qquad \alpha'_{rj} = \alpha_{rj}/\alpha_{rk} \qquad \text{for } j \neq k$$

$$\alpha'_{ik} = -\alpha_{ik}/\alpha_{rk} \qquad \text{for } i \neq r; \qquad \alpha'_{ij} = \alpha_{ij} - \frac{\alpha_{rj}\,\alpha_{ik}}{\alpha_{rk}} \qquad \text{for } i \neq r,\ j \neq k.$$

No special rules are required for the resource column; we simply substitute $j = 0$. This is the reason for employing the $j = 0$ convention to refer to the resource column even though the entries are not coefficients. The rule for $\alpha'_{ij}$ can be rewritten in the simpler form

$$\alpha'_{ij} = \alpha_{ij} + \alpha_{rj}\,\alpha'_{ik}.$$

The same transformation rules apply to the reduced tableau since this displays just the $\alpha_{ij}$s. In this case $x_{B_r}$ occurs in the $r$th row and $x_{N_k}$ in the $k$th column. The whole operation is called **pivoting** and the $r$th row and $k$th column are called, respectively, the **pivot row** and **pivot column**. The unique element in both the pivot row and the pivot column is called the **pivot.** The pivoting rules can then be expressed verbally (the easiest way to use them in hand calculation) as follows.

(1) Replace the pivot by its reciprocal.
(2) Divide the remainder of the pivot row by the pivot.
(3) Multiply the remainder of the pivot column by $-1$ and divide by the pivot.
(4) Add $\alpha'_{ik}$ (the new element in row $i$ and the pivot column – calculated in operation (3)) times the pivot row to row $i$, omitting entries in the pivot column, for all $i \neq r$.

(Adding, multiplying and dividing columns or rows means performing these operations element by element.) Rule (4) is based on the alternative formula for $\alpha'_{ij}$.

These pivoting rules may appear complicated on first reading, so we will illustrate them by pivoting in reduced tableau T1*, to make $x_2$ basic and $x_4$ non-basic. (The reader should be assured, however, that practice on a few exercises will soon make the procedure appear transparently obvious.) The working has been set out explicitly. The result agrees, of course, with our previous calculations. A further pivot in row 1 and column 2 makes $x_1$, $x_2$ and $x_5$ basic, giving a tableau T3*, which could not have been obtained in a single pivot from T1* (why?).

| | $x_4$ | $x_5$ | T2* |
|---|---|---|---|
| $x_3$ | $\frac{2}{3}$ | $1+\frac{2}{3}\times(-4)=-\frac{5}{3}$ | $1+\frac{2}{3}\times 9=7$ |
| $x_1$ | $-\frac{1}{3}$ | $0+(-\frac{1}{3})\times(-4)=\frac{4}{3}$ | $-7+(-\frac{1}{3})\times 9=-10$ |
| $x_2$ | $\frac{1}{3}$ | $-\frac{4}{3}$ | 3 |

| | $x_4$ | $x_3$ | T3* |
|---|---|---|---|
| $x_5$ | $-\frac{2}{5}$ | $-\frac{3}{5}$ | $-4\frac{1}{5}$ |
| $x_1$ | $\textcircled{\frac{1}{5}}$ | $\frac{4}{5}$ | $-4\frac{2}{5}$ |
| $x_2$ | $-\frac{1}{5}$ | $-\frac{4}{5}$ | $-2\frac{3}{5}$ |

To indicate the pivot row and column, in any tableau, we will put a ring round the pivot. Thus the pivot in T3* would lead to a new tableau with basis $\{x_2, x_4, x_5\}$ and the reader is urged to perform the pivoting operating. From now on, we will usually use the reduced tableau to perform pivoting. Indeed, we shall refer to it simply as a tableau and call the original version an **extended tableau.**

It is easy to make arithmetical slips when performing pivoting operations and so methods of checking if a mistake has been made are useful. Many methods have been proposed, but a simple procedure, requiring few extra calculations, is to choose a solution to the original canonical form and then check that it satisfies subsequent canonical forms. One simple choice is to put all the non-basic variables equal to 1 and calculate the resulting values of the basic variables. Thus, in T1* we put $x_2 = x_5 = 1$ and then we get

$$x_3 - 2 \times 1 + 1 \times 1 = 1, \qquad \text{so } x_3 = 2,$$

and, similarly, $x_1 = -8, x_4 = 10$. To check T3*, for row 1 we calculate

$$x_5 - \tfrac{2}{5}x_4 - \tfrac{3}{5}x_3 = 1 - \tfrac{2}{5} \times 10 - \tfrac{3}{5} \times 2 = -4\tfrac{1}{5}$$

which agrees with the resource column of T3*. Similarly, we can check the other rows. Of course, the check is not infallible; an incorrect tableau may still pass the test, but errors are usually picked up. One way to use the method, when a sequence of pivoting operations is being performed, is to apply the test to the final tableau and, if a mistake is detected, the test can be used in an attempt to find the first tableau which is erroneous.

This test uses the fact that any solution of a canonical form is also a solution to the new form, after pivoting. This result is a consequence of the observation that the only operations we have performed are to multiply equations by constants and add or subtract the resulting equations. Furthermore, any pivoting operation can be reversed, meaning that, if we pivot to swap $x_{B_r}$ and $x_{N_k}$ and, in the resulting tableau pivot again to make $x_{N_k}$ non-basic and $x_{B_r}$ basic once more, then we restore the previous basis and so return to the original tableau. Therefore, any solution of the transformed canonical form also solves the original canonical form. Putting these results together, we see that the set of solutions is not altered by pivoting. Pivoting merely changes the appearance of the equations, but the set of solutions remains the same. This obviously applies also after a sequence of pivots.

One particular solution which is suggested by any canonical form is that in which the non-basic variables are set to zero. The basic variables will then be equal to the resource column of the tableau (why?). This solution is called a **basic** solution. The basic solutions for tableaux T1*, T2* and T3* are

$$\begin{aligned}(x_1, x_2, x_3, x_4, x_5) &= (-7, 0, 1, 9, 0) \text{ in T1}^* \\ &= (-10, 3, 7, 0, 0) \text{ in T2}^* \\ &= (-4\tfrac{2}{5}, -2\tfrac{3}{5}, 0, 0, -4\tfrac{1}{5}) \text{ in T3}^*.\end{aligned}$$

We will see, in subsequent chapters, that basic solutions play a crucial role in linear programming.

## 2.3 BASIC FEASIBLE SOLUTIONS

Our discussion of systems of equations has, so far, overlooked the requirement that variables should be non-negative. In particular, a basic solution of the constraints is only feasible if all the variables are non-negative, in which case we will call it a **Basic Feasible Solution** (BFS). It turns out that BFSs are closely related to vertices and to illustrate this we shall consider the set of constraints:

$$\begin{aligned} 3x_1 - 2x_2 &\leqslant 6 \\ x_1 - 2x_2 &\leqslant 1 \\ -3x_1 + 2x_2 &\leqslant 3 \\ x_1, x_2 &\geqslant 0 \end{aligned}$$

which can be rewritten, by introducing slack variables, as

$$\begin{aligned} s_1 \qquad\qquad\qquad &+3x_1 - 2x_2 = 6 \\ s_2 \qquad\quad &+ x_1 - 2x_2 = 1 \\ s_3 \quad &-3x_1 + 2x_2 = 3 \\ s_1, s_2, s_3, x_1, x_2 &\geqslant 0. \end{aligned}$$

These equations are in canonical form with $s_1$, $s_2$ and $s_3$ basic and the corresponding tableau is T4. The basic solution is $(s_1, s_2, s_3, x_1, x_2) = (6, 1, 3, 0, 0)$.

| | ↓ $x_1$ | $x_2$ | T4 |
|---|---|---|---|
| $s_1$ | 3 | −2 | 6 |
| → $s_2$ | (1) | −2 | 1 |
| $s_3$ | −3 | 2 | 3 |

By choosing any of the six possible pivots in T4, we can obtain six new tableaux and, therefore, six new basic solutions. In each of these new tableaux we look for pivots which lead to tableaux and basic solutions not already encountered. In this way, we can generate all possible tableaux and basic solutions (cf. Exercise 2). The reader should check that the basic solutions are as follows:

$$\begin{aligned} (s_1, s_2, s_3, x_1, x_2) &= (6, 1, 3, 0, 0)\text{: A} \\ &= (9, 2, 0, -1, 0)\text{: B} \\ &= (3, 0, 6, 1, 0)\text{: C} \\ &= (0, -1, 9, 2, 0)\text{: D} \\ &= (9, 4, 0, 0, 1\tfrac{1}{2})\text{: E} \\ &= (9, 0, 0, -2, -1\tfrac{1}{2})\text{: F} \end{aligned}$$

$$= (0, 0, 9, 2\tfrac{1}{2}, \tfrac{3}{4}): \text{G}$$
$$= (5, 0, 4, 0, -\tfrac{1}{2}): \text{H}$$
$$= (0, -5, 9, 0, -3): \text{J}.$$

For any of these basic solutions, the basic variables can be identified as having non-zero values and the non-basic variables as having zero values. Furthermore, pivoting is possible between the tableaux corresponding to a pair of basic solutions, if and only if the solutions have exactly one non-basic variable in common (why?). Thus from D it is possible to pivot to A, B, C, G and J but not to E, F or H.

The inequalities are displayed graphically in Fig. 2.1. The constraints have been labelled with the corresponding slack variables. The values of $x_1$ and $x_2$ in each of the basic solutions in our list specify a point in the figure and each of these points has been labelled with the appropriate basic solution. We see that basic solutions specify points at the intersections of a pair of lines (constraints or axes). This is not surprising since each line corresponds to setting one variable to zero and a basic solution has two variables set to zero. Furthermore, we see that pivoting corresponds to moving along a line from one intersection to any other intersection on that line. Thus, from C, we can make $s_2$ basic (leaving $x_2 = 0$), and pivot to A, B or D, or we can make $x_2$ basic and pivot to F, G or H. Which of A, B, D or F, G, H we arrive at is determined by the choice of basic variable which is to become non-basic.

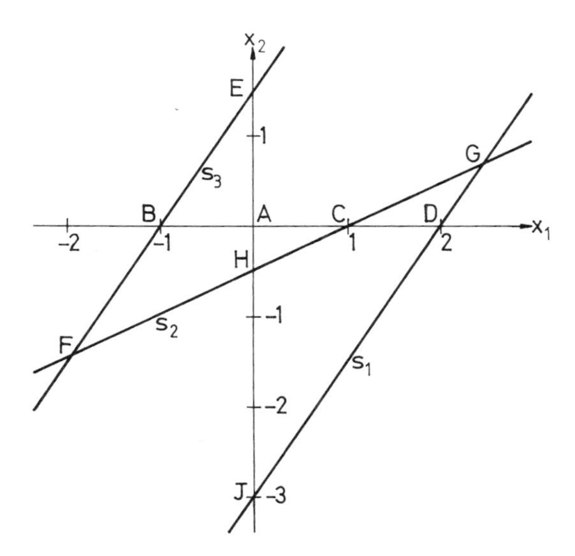


Figure 2.1 Basic solutions: graphical interpretation.

We have seen that basic solutions correspond to intersections. It follows that BFSs correspond to vertices and this can be seen in the figure, the vertices being A, C, E and G. This correspondence (which can be proved true, in general) suggests, in the light of our observations in Section 1.3 on the optimality of vertices, that in seeking optimal solutions of LPs we can restrict the search to BFSs. In the next chapter we will discover that this suggestion is justified.

## 2.4 MAINTAINING PRIMAL FEASIBILITY – PIVOT ROW SELECTION

The importance of BFSs prompts the question: when pivoting from a BFS how can we ensure that the transformed tableau also corresponds to a BFS?

The basic solution corresponding to any tableau satisfies

$$x_{B_i} = \alpha_{io} \quad \text{and} \quad x_{N_j} = 0,$$

so this solution is feasible if $\alpha_{io} \geqslant 0$ for all $i$ and we will call the tableau **primal feasible.** Thus, our question can be re-phrased: how can we ensure that primal feasibility is maintained when pivoting? In Fig. 2.1 we see that from A it is possible to pivot along one axis to B, C or D and along the other to E, H or J but, in each case, only one of the new intersections on each line is feasible. This suggests that, if the non-basic variable which must enter the basis is specified, a correct choice of variable to leave the basis is needed to maintain primal feasibility. Suppose, then, that $x_{N_k}$ is to be made basic. We must choose $r$ so that, after pivoting in row $r$ and column $k$, we have $\alpha'_{io} \geqslant 0$ for all $i$, in the trans-tableau. To examine the restrictions such a requirement places on the choice of $r$, we will assume, for the moment, that $\alpha_{ro} > 0$. Then, the transformation rules for pivoting tell us that

$$\alpha'_{ro} = \alpha_{ro}/\alpha_{rk}$$

and requiring $\alpha'_{ro} \geqslant 0$ forces us to choose $\alpha_{rk} > 0$. Furthermore, for $i \neq r$,

$$\alpha'_{io} = \alpha_{io} - \frac{\alpha_{ik}\,\alpha_{ro}}{\alpha_{rk}} = \alpha_{io} - \alpha_{ik}\,\alpha'_{ro}$$

and, if $\alpha_{ik} \leqslant 0$, we have, since $\alpha'_{ro} \geqslant 0$,

$$\alpha'_{io} \geqslant \alpha_{io} \geqslant 0,$$

so non-negativity of $\alpha'_{io}$ is guaranteed. However, if $\alpha_{ik} > 0$, the requirement

$$\alpha_{io} - \frac{\alpha_{ik}\,\alpha_{ro}}{\alpha_{rk}} \geqslant 0$$

can be divided by the positive number $\alpha_{ik}$ to obtain

$$\frac{\alpha_{io}}{\alpha_{ik}} \geqslant \frac{\alpha_{ro}}{\alpha_{rk}}.$$

We can ensure that this inequality is satisfied by choosing $r$ to minimise the left-hand ratio, that is, we choose $r$ to satisfy

$$\alpha_{rk} > 0$$

$$\frac{\alpha_{ro}}{\alpha_{rk}} = \min_{i \geqslant 1} \left\{ \frac{\alpha_{io}}{\alpha_{ik}} \,\middle|\, \alpha_{ik} > 0 \right\}$$

where the last expression is shorthand for taking the minimum over all $i$ for which $\alpha_{ik} > 0$.

We shall call this rule for selecting $r$ the **Pivot Row Selection** (PRS) rule. It may be expressed verbally by saying that, once the pivot column is specified, we ignore rows with a non-positive entry in that column. Of the remaining rows we select one in which the ratio of the entry in the resource column to that in the pivot column is minimised. For example, in T4, if the first column is chosen (make $x_1$, basic), then we ignore row 3 (negative entry in pivot column) and, since

$$\tfrac{1}{1} < \tfrac{6}{3},$$

select row 2. Pivoting as indicated in T4 gives the tableau T5, corresponding to basic solution C.

| | $s_2$ | ↓ $x_2$ | T5 |
|---|---|---|---|
| → $s_1$ | −3 | (4) | 3 |
| $x_1$ | 1 | −2 | 1 |
| $s_3$ | 3 | −4 | 6 |

In T5, if we wish to make $x_2$ basic, the only positive entry in the $x_2$– column, and, therefore, the pivot, is in row 1. Pivoting, we get T6, corresponding to G.

| | $s_2$ | $s_1$ | T6 |
|---|---|---|---|
| $x_2$ | $-\frac{3}{4}$ | $\frac{1}{4}$ | $\frac{3}{4}$ |
| $x_1$ | $-\frac{1}{2}$ | $\frac{1}{2}$ | $2\frac{1}{2}$ |
| $s_3$ | 0 | 1 | 9 |

We observe that, after both pivots, the resource columns are non-negative, as we desired. Indeed, if an element in the resource column becomes negative, a mistake, either in the use of the PRS rule or in pivoting, must have been made.

To derive the PRS rule, we made the temporary assumption that $\alpha_{ro} > 0$. However, consider the tableau T7 which is **degenerate** (has at least one zero in

| | $x_4$ | $x_2$ | T7 |
|---|---|---|---|
| $x_1$ | $\frac{1}{2}$ | $-\frac{1}{2}$ | $\frac{1}{2}$ |
| $x_3$ | $-\frac{3}{2}$ | 2 | 0 |
| $x_5$ | $\frac{3}{2}$ | $\frac{1}{2}$ | 1 |

in the resource column). If we choose to make $x_4$ basic, the PRS rule decrees that we must make $x_5$ non-basic. However, if we ignore the negativity of the pivot and make $x_3$ basic, the reader can verify that the resource column does not change and so the resulting tableau is primal feasible. To see that it is always true that, if $\alpha_{r0} = 0$, pivoting in row $r$ does not alter the resource column, note that

$$\alpha'_{r0} = \alpha_{r0}/\alpha_{rk} = 0 = \alpha_{r0}$$

$$\alpha'_{i0} = \alpha_{i0} - \frac{\alpha_{ik}\,\alpha_{r0}}{\alpha_{rk}} = \alpha_{i0} \qquad \text{for } i \neq r$$

provided only that $\alpha_{rk} \neq 0$. However, this result also means that, if the PRS rule *is* used, primal feasibility will still be maintained even if $\alpha_{r0} = 0$. In other words, whilst we *must* use the PRS rule for non-degenerate tableaux, we can also use it for degenerate tableaux. Thus, whenever primal feasibility is to be maintained, we will always use the PRS rule. Note that this forces us to pivot in a row $r$ where $\alpha_{r0} = 0$, if $\alpha_{rk} > 0$. For example, the only pivot maintaining primal feasibility in the $x_2$–column of T7 is in the $x_3$–row.

The argument so far may suggest that any column will contain a pivot maintaining primal feasibility. This is not necessarily true for, if we try to make $s_2$ basic in T6, the PRS rule breaks down as there is no positive entry in the pivot column. In Fig. 2.1, we see that making $s_2$ basic corresponds to moving from G along the line JDG, but away from D. This takes us along an unbounded edge of the feasible region, which explains why we will not encounter a new vertex or BFS. To investigate this in general, suppose that we wish to make $x_{N_k}$ basic but find $\alpha_{ik} \leqslant 0$ for all $i$. Then, if we put

$$x_{N_k} = \lambda$$

$$x_{N_j} = 0 \qquad \text{for } j \neq k$$

$$x_{B_i} = \alpha_{i0} - \lambda\alpha_{ik} \qquad \text{for all } i,$$

we see that this solution satisfies the canonical equations:

$$x_{B_i} + \sum_j \alpha_{ij}\, x_{N_j} = \alpha_{i0} \qquad \text{for all } i.$$

Furthermore, if $\lambda \geqslant 0$, we obviously have $x_{N_k} \geqslant 0$ and $x_{B_i} \geqslant \alpha_{i0} \geqslant 0$, for all $i$, so that the solution is feasible for any $\lambda \geqslant 0$. Since $\lambda$ can be arbitrarily large, the feasible region must be unbounded. In the example, we obtain from T6 the solution, for $\lambda \geqslant 0$,

$$(s_1, s_2, s_3, x_1, x_2) = (0, \lambda, 9, 2\tfrac{1}{2} + \tfrac{1}{2}\lambda, \tfrac{3}{4} + \tfrac{3}{4}\lambda),$$

the final two components of which define, in parametrised form, the unbounded edge of the feasible region starting at G.

In the examples we have looked at so far, the PRS rule has led to a unique choice of pivot row. This need not always be the case, as the tableau T8 shows.

| | $s_2$ | $x_1$ | T8 |
|---|---|---|---|
| $x_2$ | $\frac{1}{4}$ | $\frac{7}{4}$ | 1 |
| $s_1$ | $\frac{3}{4}$ | $-\frac{3}{4}$ | $2\frac{1}{4}$ |
| $s_3$ | $\frac{5}{4}$ | $\frac{1}{4}$ | $3\frac{3}{4}$ |

If we try to make $s_2$ basic, the ratios in the second and third rows both minimise $\alpha_{i0}/\alpha_{ik}$ and so either could be chosen as pivot row. Each choice would lead to a (different) primal feasible tableau. However, for either choice the new BFS will be degenerate. This is a general result (Exercise 7). The PRS rule specifies no way of breaking such ties. One possible method will be described in Section 3.6, but until then we will assume an arbitrary choice is made when ties occur. This will lead to no difficulties in practice.

## EXERCISES

1. What are the basic and non-basic variables in the following set of equations in canonical form?

$$\begin{aligned} s_1 \qquad\qquad\quad & + x_1 + 3x_2 - x_3 = 2 \\ s_2 \qquad & +2x_1 - x_2 + 4x_3 = 5 \\ s_3 & -2x_1 + 3x_2 + x_3 = -2. \end{aligned}$$

Using as few pivoting operations as possible, express the set of equations in canonical form with basis $\{s_3, x_1, x_3\}$. Do this exercise using both extended and reduced tableaux. Use the checking procedure of Section 2.2 to test your results.

2. By writing the equations

$$x_1 + 3x_2 + 2x_3 - 2x_4 = 1$$
$$-3x_1 - 6x_2 - x_3 + 4x_4 = 5$$

as

$$x_1 + 2x_3 = 1 - 3x_2 + 2x_4$$
$$-3x_1 - x_3 = 5 + 6x_2 - 4x_4$$

and 'solving' for $x_1$ and $x_3$ in terms of RHSs involving $x_2$ and $x_4$, find a canonical form of these equations with basis $\{x_1, x_3\}$. Using pivoting in the reduced tableau, find *all* basic solutions of the system of equations.

3. (Continuation of Exercise 2, assuming a knowledge of determinants.)
For each pair of variables, evaluate the determinant of the coefficients of these variables in the equations in Exercise 2. For example, for $x_2$ and $x_3$ you would have to evaluate

$$\begin{vmatrix} 3 & 2 \\ -6 & -1 \end{vmatrix}.$$

Compare the results of Exercises 2 and 3 and comment.

4. If, in any reduced tableau, you pivot in row $r$ and column $k$ and, in the subsequent tableau, again pivot in row $r$ and column $k$, show, using the transformation rules directly, that the result is a return to the original tableau.

5. Assuming that the PRS rule is used, write down each of the tableaux that are reached in one pivot from the following, i.e. try making each non-basic variable basic.

(i)

| | $x_5$ | $x_2$ | $x_4$ | |
|---|---|---|---|---|
| $x_3$ | $\frac{3}{2}$ | $\frac{3}{4}$ | $\frac{13}{4}$ | $3\frac{1}{4}$ |
| $x_1$ | $\frac{1}{2}$ | $-\frac{5}{4}$ | $\frac{5}{4}$ | $1\frac{1}{4}$ |
| $x_6$ | $\frac{1}{2}$ | $-\frac{3}{4}$ | $\frac{7}{4}$ | $2\frac{3}{4}$ |

(ii)

| | $x_1$ | $x_2$ | $x_3$ | |
|---|---|---|---|---|
| $s_1$ | 3 | 7 | −2 | 4 |
| $s_2$ | −4 | 1 | −2 | 0 |
| $s_3$ | 5 | −3 | 0 | 6 |

Are there any other tableaux which define a BFS and can be reached in one pivot for example (ii)? Write down, in parametrised form, an unbounded family of solutions to the equations represented by the second tableau.

6. Write each of the following systems of inequalities as a set of equalities, using slack variables. By pivoting, find all BFSs in each problem and check your answers graphically.

(i)
$$\begin{aligned} 3x_1 + x_2 &\leqslant 9 \\ x_1 + x_2 &\leqslant 4 \\ x_1 + 4x_2 &\leqslant 12 \\ x_1, x_2 &\geqslant 0. \end{aligned}$$

(ii)
$$\begin{aligned} -3x_1 + x_2 &\leqslant 3 \\ -x_1 + x_2 &\leqslant 2 \\ x_1 - x_2 &\leqslant 3 \\ x_1 - 4x_2 &\leqslant 4 \\ x_1, x_2 &\geqslant 0. \end{aligned}$$

You may assume that every BFS may be reached by pivoting, using the PRS rule, from the initial tableau (but not necessarily in one pivot).

7. Prove that if the pivot row selection rule leads to a tie (choice of rows) in some column then any tableau obtained by pivoting in that column will be degenerate if it defines a BFS.

Deduce that, provided degeneracy is not encountered, if the pivot row selection rule specifies row $r$ when pivoting in column $k$ the same will be true in the transformed tableau, i.e. if, after pivoting in column $k$ it is desired to pivot again in column $k$, the pivot row selection rule will determine $r$ as pivot row.

CHAPTER 3

# The Simplex Method

## 3.1 THE OBJECTIVE FUNCTION AND THE TABLEAU

In our discussion of BFSs we have been concerned, so far, only with the constraints of LPs. In this chapter, we will see how the objective function can be included in tableaux. This will lead to a procedure for solving LPs known as the **simplex method.**

We will start by looking at the example

$$\text{P1:}\quad \begin{array}{lrl} \text{maximise} & 3x_1 - 4x_2 & \quad (=z) \\ \text{subject to} & 3x_1 - 2x_2 \leqslant 6 & \\ & x_1 - 2x_2 \leqslant 1 & \\ & -3x_1 + 2x_2 \leqslant 3 & \\ & x_1, x_2 \geqslant 0, & \end{array}$$

in which the constraints are identical with the example of Section 2.3. The objective function can be incorporated into the canonical form by writing it as

$$z - 3x_1 + 4x_2 = 0$$

and treating $z$ as an extra basic variable. In tableau form, this can be written:

| P1 | $x_1$ | $x_2$ | T1 |
|---|---|---|---|
| $s_1$ | 3 | −2 | 6 |
| $s_2$ | (1) | −2 | 1 |
| $s_3$ | −3 | 2 | 3 |
| $z$ | −3 | 4 | 0 |

The special nature of the $z$-row, or **objective row**, is indicated by always making it the bottom row in a tableau, as shown. The reader should note that the coefficients of the variables in the objective function appear multiplied by $-1$ in the tableau and be aware of the reason why.

In Chapter 2 we discussed evidence that, in seeking an optimal solution, we can look for an optimal BFS. This suggests that we try pivoting in P1/T1 in an attempt to find a BFS which improves the objective function value. If we make $x_1$ basic, $x_1$ will change in value from zero to some positive value, say $\mu > 0$. Consequently, $z$ will increase from zero to $3\mu$. If we make $x_2$ basic, it will become positive and $z$ will decrease. We therefore, choose the first option. When we pivot to make $x_1$ basic, the pivot row selection rule requires $s_2$ to leave the basis. Making the indicated pivot gives P1/T2. Since we regard $z$ as an extra basic variable, we can transform the objective row by the same rules as the other rows, when pivoting. In this way we always have $z$ expressed in terms of the non-basic variables. Of course, we never pivot in the objective row; the PRS rule applies only to the other rows.

| P1 | $s_2$ | $x_2$ | T2 |
|---|---|---|---|
| $s_1$ | −3 | (4) | 3 |
| $x_1$ | 1 | −2 | 1 |
| $s_3$ | 3 | −4 | 6 |
| $z$ | 3 | −2 | 3 |

In particular, in P1/T2, the objective row represents the equation

$$z + 3s_2 - 2x_2 = 3,$$

which shows that increasing $s_2$ decreases $z$ (not surprisingly, since pivoting in the $s_2$-column would take us back to P1/T1) and increasing $x_2$ increases $z$. We therefore make $x_2$ basic by pivoting as indicated. The result is tableau P1/T3. The objective row represents

$$z + 1\tfrac{1}{2}s_2 + \tfrac{1}{2}s_1 = 4\tfrac{1}{2}$$

and increasing either non-basic variable decreases $z$. Since no improvement is possible, this suggests we have an optimal solution.

The procedure described in Section 2.2 for checking on mistakes in pivoting extends to the objective row in the obvious way. For example, if $x_1 = x_2 = 1$ in P1/T1 we must have $s_1 = 5$, $s_2 = 2$, $s_3 = 4$ and $z = -1$. Then to check the objective row of P1/T3, we verify that $-1 + 1\tfrac{1}{2} \times 2 + \tfrac{1}{2} \times 5 = 4\tfrac{1}{2}$.

In our solution of P1 we pivoted in each case in a column with a negative entry in the objective row. When no such entry could be found, we concluded that the BFS was optimal. We shall now generalise this procedure and

| P1 | $s_2$ | $s_1$ | T3 |
|---|---|---|---|
| $x_2$ | $-\frac{3}{4}$ | $\frac{1}{4}$ | $\frac{3}{4}$ |
| $x_1$ | $-\frac{1}{2}$ | $\frac{1}{2}$ | $2\frac{1}{2}$ |
| $s_3$ | 0 | 1 | 9 |
| $z$ | $1\frac{1}{2}$ | $\frac{1}{2}$ | $4\frac{1}{2}$ |

demonstrate its validity. Firstly, we shall write the equation represented by the objective row of the tableau as

$$z + \sum_j \alpha_{0j} x_{N_j} = \alpha_{00}. \tag{3.1}$$

In other words we are viewing the row and its equation as a special case, $i = 0$, of the general form of the canonical equation. This involves making the identification $x_{B_0} = z$, which is permissible, since $z$ never becomes non-basic. The corresponding BFS has $x_{N_j} = 0$ for all $j$ which means that $\alpha_{00}$ is the value of the objective function at that solution. When pivoting, the objective row is treated like any other row. Equivalently, we can put $i = 0$ in the transformation rules for pivoting. In particular, we have

$$\alpha_{00}' = \alpha_{00} - \frac{\alpha_{0k}\,\alpha_{r0}}{\alpha_{rk}}.$$

Assuming non-degeneracy for the moment, we must have $\alpha_{r0} > 0$, $\alpha_{rk} > 0$ and thus, if we choose $k$ so that $\alpha_{0k} < 0$, then $\alpha_{00}' > \alpha_{00}$. This says that the objective function increases, if we pivot in a column which has a negative entry in the objective row. However, if there is no such column, which means that $\alpha_{0j} \geqslant 0$ for all $j$, then, since any feasible solution of the problem has $x_{N_j} \geqslant 0$ for all $j$, the objective function value $z$, of any feasible solution, must satisfy

$$z = \alpha_{00} - \sum_j \alpha_{0j} x_{N_j} \leqslant \alpha_{00}.$$

This says that no other feasible solution has a greater objective function value than the current BFS, which has value $\alpha_{00}$. Equivalently, we have located an optimal solution.

These arguments indicate that we should proceed by selecting a $k \geqslant 1$ with $\alpha_{0k} < 0$ and pivoting in column $k$. This is repeated, improving the objective function each time, until a tableau is reached in which no such $k$ can be found and which can be declared optimal. In essence, this is the simplex method but, in general $\alpha_{0k}$ will be negative for more than one $k$ and to complete our description we need a more precise rule for selecting a pivot column. One such rule is motivated by observing that if $x_{N_k}$ increases by $\mu$, then $z$ increases by $(-\alpha_{0k})\mu$, assuming the remaining non-basic variables stay at zero. Thus, the maximum rate of increase of $z$ is obtained by choosing $k$ to maximise $-\alpha_{0k}$ or, equivalently, we take $k \geqslant 1$ to satisfy

$$\alpha_{0k} = \min_{j \geqslant 1} \alpha_{0j}.$$

Ties for the minimum may be broken arbitrarily. We shall have more to say about pivot column selection in Section 4.4 and the exercises in Chapter 4, but in the sequel we shall assume that the above rule is being used, unless specific indications are given to the contrary.

We shall now solve the following example by the simplex method

$$\begin{array}{llrl} \text{P2:} & \text{minimise} & x_1 - 2x_2 - 3x_3 & (= -z) \\ & \text{subject to} & 3x_1 - x_2 - x_3 \leqslant 3 & \\ & & -x_1 - x_2 + 2x_3 \leqslant 5 & \\ & & -x_1 + 2x_2 - x_3 \leqslant 3 & \\ & & x_1, x_2, x_3 \geqslant 0. & \end{array}$$

As explained in Chapter 1, we convert this to the problem of maximising $z = -x_1 + 2x_2 + 3x_3$. So the initial tableau is P2/T1 and we must take care to remember that we started with a minimisation problem.

| P2 | $x_1$ | $x_2$ | $x_3$ | T1 |
|---|---|---|---|---|
| $s_1$ | 3 | $-1$ | $-1$ | 3 |
| $s_2$ | $-1$ | $-1$ | (2) | 5 |
| $s_3$ | $-1$ | 2 | $-1$ | 3 |
| $z$ | 1 | $-2$ | $-3$ | 0 |

Both the $x_2$ and $x_3$ columns have negative entries in the objective row. Since $-3 < -2$ we pivot in the third ($x_3$) column. This means making $s_2$ non-basic and the next tableau is P2/T2. In P2/T2, the most negative element in the objective bottom row occurs in the second column. This leads to the indicated pivot. Following the same rules, we obtain two more tableaux. Since it has a

| P2 | $x_1$ | $x_2$ | $s_2$ | T2 |
|---|---|---|---|---|
| $s_1$ | $\frac{5}{2}$ | $-\frac{3}{2}$ | $\frac{1}{2}$ | $5\frac{1}{2}$ |
| $x_3$ | $-\frac{1}{2}$ | $-\frac{1}{2}$ | $\frac{1}{2}$ | $2\frac{1}{2}$ |
| $s_3$ | $-\frac{3}{2}$ | ($\frac{3}{2}$) | $\frac{1}{2}$ | $5\frac{1}{2}$ |
| $z$ | $-\frac{1}{2}$ | $-3\frac{1}{2}$ | $1\frac{1}{2}$ | $7\frac{1}{2}$ |

non-negative objective row, P2/T4 is an optimal tableau. The optimal solution is $(x_1, x_2, x_3) = (11, 14\frac{2}{3}, 15\frac{1}{3})$ and the optimal objective function value $z = 64\frac{1}{3}$, so that the value of the original objective function (the one to be minimised) is $-64\frac{1}{3}$.

| P2 | $x_1$ | $s_3$ | $s_2$ | T3 |
|---|---|---|---|---|
| $s_1$ | (1) | 1 | 1 | 11 |
| $x_3$ | −1 | $\frac{1}{3}$ | $\frac{2}{3}$ | $4\frac{1}{3}$ |
| $x_2$ | −1 | $\frac{2}{3}$ | $\frac{1}{3}$ | $3\frac{2}{3}$ |
| $z$ | −4 | $2\frac{1}{3}$ | $2\frac{2}{3}$ | $20\frac{1}{3}$ |

| P2 | $s_1$ | $s_3$ | $s_2$ | T4 |
|---|---|---|---|---|
| $x_1$ | 1 | 1 | 1 | 11 |
| $x_3$ | 1 | $\frac{4}{3}$ | $\frac{5}{3}$ | $15\frac{1}{3}$ |
| $x_2$ | 1 | $\frac{5}{3}$ | $\frac{4}{3}$ | $14\frac{2}{3}$ |
| $z$ | 4 | $6\frac{1}{3}$ | $6\frac{2}{3}$ | $64\frac{1}{3}$ |

## 3.2 FINDING AN INITIAL PRIMAL FEASIBLE TABLEAU – EXAMPLES

The simplex method employs pivoting between primal feasible tableaux and, therefore, in order to start the method going, an initial primal feasible tableau is required. This means that, once an LP has been converted to standard equality form, we must select a basis for which the corresponding basic solution is feasible. In the examples solved in Section 3.1, the constraints automatically satisfied this requirement, with the slack variables as the basis. Indeed, we did not even explicitly write down the standard equality forms of the LPs. Unfortunately we cannot always expect to be so fortunate, as problem P3 shows.

$$\text{P3:}\quad \left.\begin{array}{ll} \text{minimise} & 3x_1 + 4x_2 + 5x_3 \; (= -z) \\ \text{subject to} & x_1 + x_2 - 2x_3 \leqslant -1 \\ & 2x_1 + x_2 - x_3 \geqslant -2 \\ & x_1 - x_2 + 3x_3 = 3 \\ & 2x_1 - x_2 - x_3 \leqslant 2 \\ & x_1, x_2, x_3 \geqslant 0 \end{array}\right\}$$

$$\left\{\begin{array}{lll} \text{maximise} & -3x_1 - 4x_2 - 5x_3 & (= z) \\ \text{subject to} & -x_1 - x_2 + 2x_3 - s_1 & = 1 \\ & -2x_1 - x_2 + x_3 + s_2 & = 2 \\ & x_1 - x_2 + 3x_3 & = 3 \\ & 2x_1 - x_2 - x_3 + s_4 & = 2 \\ & x_1, x_2, x_3, s_1, s_2, s_4 \geqslant 0. & \end{array}\right.$$

We have written out the standard equality form and multiplied constraints by −1 where necessary to ensure that all RHSs are non-negative.

Taking $s_2$ and $s_4$ basic in the second and fourth equations is satisfactory, since they will then have positive values in the corresponding basic solution. However, if we took $s_1$ basic in the first equation, we would have $s_1 = -1$ in the basic solution, which is then infeasible. Even worse, there is no obvious basic variable in the third equation. However, we shall see that the simplex method itself can be used to resolve our dilemma and this is the reason why we postponed the problem of initialisation until we had described how simplex iterations were performed.

To apply the simplex method to the problem of finding an initial tableau, we start by adding additional, **artificial variables** to the left-hand sides of those constraints lacking a suitable basic variable (the second and fourth in our example). In P3 the result is

$$\begin{aligned}
-x_1 - x_2 + 2x_3 - s_1 + a_1 &= 1\\
-2x_1 + x_2 + 3x_3 + s_2 &= 2\\
x_1 - x_2 + 3x_3 + a_3 &= 3\\
2x_1 - x_2 - x_3 + s_4 &= 2,
\end{aligned}$$

which is in canonical form with respect to $a_1, s_2, a_3, s_4$, and the basic solution is feasible. However, this has been achieved at the cost of introducing extra variables. Nevertheless, if we can pivot to make both the artificial variables non-basic, then all the basic variables in that tableau must be original (non-artificial) variables. Consequently, setting the artificial variables to zero will give a primal feasible tableau containing only the original variables. To drive the artificial variables out of the basis, we can use the simplex method to maximise $-a_1-a_3(\equiv z^{\mathrm{I}})$. Provided the LP is feasible, the optimal solution must have $a_1 = a_3 = 0$ and, assuming non-degeneracy (an assumption we shall maintain throughout this section, but will later relax), this means $a_1$ and $a_3$ must be non-basic in the optimal tableau for $z^{\mathrm{I}}$. Setting non-basic variables (such as $a_1$ and $a_3$ at the optimum) to zero is equivalent to dropping the corresponding columns.

Before we can start the maximisation of $z^{\mathrm{I}}$ we must express it in terms of the non-basic variables $x_1$, $x_2$, $x_3$, $s_1$ in the canonical form above so that it can be entered in the tableau. Multiplying the first and third equations by $-1$ and adding the results, we obtain

$$z^{\mathrm{I}} + 2x_2 - 5x_3 + s_1 = -4$$

which appears as the objective row of P3/T1.

We now proceed to maximise $z^{\mathrm{I}}$. The first pivot is indicated in P3/T1. After two pivots we reach the optimal tableau P3/T2. Since both artificial variables are non-basic in P3/T2, we drop the columns indicated by * and we are left with a primal feasible tableau but with no objective row. The missing row expresses $z$ in terms of the non-basic variables $x_2$ and $s_1$ and can, therefore, be written

| P3 | $x_1$ | $x_2$ | $x_3$ | $s_1$ | T1 |
|---|---|---|---|---|---|
| $a_1$ | −1 | −1 | (2) | −1 | 1 |
| $s_2$ | −2 | 1 | 3 | 0 | 2 |
| $a_3$ | 1 | −1 | 3 | 0 | 3 |
| $s_4$ | 2 | −1 | −1 | 0 | 2 |
| $z^{\mathrm{I}}$ | 0 | 2 | −5 | 1 | −4 |

| P3 | * $a_3$ | $x_2$ | * $a_1$ | $s_1$ | T2 |
|---|---|---|---|---|---|
| $x_3$ | $\frac{1}{5}$ | $-\frac{2}{5}$ | $\frac{1}{5}$ | $-\frac{1}{5}$ | $\frac{4}{5}$ |
| $s_2$ | $\frac{3}{5}$ | $-\frac{1}{5}$ | $-\frac{7}{5}$ | $\frac{7}{5}$ | $2\frac{2}{5}$ |
| $x_1$ | $\frac{2}{5}$ | $\frac{1}{5}$ | $\frac{3}{5}$ | ($\frac{3}{5}$) | $\frac{3}{5}$ |
| $s_4$ | $-\frac{3}{5}$ | $-\frac{9}{5}$ | $\frac{7}{5}$ | $-\frac{7}{5}$ | $1\frac{3}{5}$ |
| $z^{\mathrm{I}}$ | 1 | 0 | 1 | 0 | 0 |
| $z$ | | $5\frac{2}{5}$ | | $-\frac{4}{5}$ | $-5\frac{4}{5}$ |

down by calculating the coefficients of this expression. We find

$$
\begin{aligned}
z &= -3x_1 - 4x_2 - 5x_3 \\
&= -3(\tfrac{3}{5} - \tfrac{1}{5}x_2 - \tfrac{3}{5}s_1) - 4x_2 - 5(\tfrac{4}{5} + \tfrac{2}{5}x_2 + \tfrac{1}{5}s_1)
\end{aligned}
$$

so

$$z + 5\tfrac{2}{5}x_2 - \tfrac{4}{5}s_1 = -5\tfrac{4}{5}$$

which has been written as an extra row in P3/T2, omitting entries in the discarded columns. The resulting pivot is indicated in the tableau, and pivoting gives P3/T3, which is optimal, indicating that $(x_1, x_2, x_3) = (0, 0, 1)$ is optimal in P3 with objective function value $-z = 5$.

| P3 | $x_2$ | $x_1$ | T3 |
|---|---|---|---|
| $x_3$ | $-\frac{1}{3}$ | $\frac{1}{3}$ | 1 |
| $s_2$ | $-\frac{2}{3}$ | $\frac{7}{3}$ | 1 |
| $s_1$ | $\frac{1}{3}$ | $\frac{5}{3}$ | 1 |
| $s_4$ | $-\frac{4}{3}$ | $\frac{7}{3}$ | 3 |
| $z$ | $5\frac{2}{3}$ | $1\frac{1}{3}$ | −5 |

Clearly, the feasibility of an LP implies that the maximum value of $z^{\mathrm{I}}$ must be 0. Equivalently, if we find $z^{\mathrm{I}}$ maximised at some negative value, we may conclude that the problem is infeasible. This is illustrated in the following example,

$$\text{P4:}\quad \text{maximise}\quad 2x_1 - x_2 + x_3 - x_4$$

$$\text{subject to}\quad 2x_1 - x_2 + 5x_3 - 3x_4 = 3 \qquad \text{(I)}$$

$$2x_1 + x_2 + 4x_3 - 2x_4 = 1 \qquad \text{(II)}$$

$$x_1, x_2, x_3, x_4 \geqslant 0,$$

for which the first tableau is P4/T1. Pivoting, we get the tableau P4/T2, which is optimal but with $z^{\mathrm{I}} < 0$. This means that P4 is not feasible, which can also be seen directly by noting that $2 \times \text{(I)} - 3 \times \text{(II)}$ gives

$$-2x_1 - 5x_2 - 2x_3 = 3$$

which cannot be satisfied in non-negative variables.

| P4 | $x_1$ | $x_2$ | $x_3$ | $x_4$ | T1 |
|---|---|---|---|---|---|
| $a_1$ | 2 | −1 | 5 | −3 | 3 |
| $a_2$ | 2 | 1 | (4) | −2 | 1 |
| $z^{\mathrm{I}}$ | −4 | 0 | −9 | 5 | −4 |

| P4 | $x_1$ | $x_2$ | $a_2$ | $x_4$ | T2 |
|---|---|---|---|---|---|
| $a_1$ | $-\frac{1}{2}$ | $-\frac{9}{4}$ | $-\frac{5}{4}$ | $-\frac{1}{2}$ | $1\frac{3}{4}$ |
| $x_3$ | $\frac{1}{2}$ | $\frac{1}{4}$ | $\frac{1}{4}$ | $-\frac{1}{2}$ | $\frac{1}{4}$ |
| $z^{\mathrm{I}}$ | $\frac{1}{2}$ | $2\frac{1}{4}$ | $2\frac{1}{4}$ | $\frac{1}{2}$ | $-1\frac{3}{4}$ |

## 3.3 THE TWO-PHASE SIMPLEX METHOD

The method outlined in the previous section divides the solution of an LP into two phases. In Phase I we obtain an initial primal feasible tableau, if one exists, using an artificial objective function and in Phase II the original objective function is maximised (or minimised). In this section we will amplify and extend Phase I.

In our solution of P3 we introduced an artificial variable $a_1$ into the first constraint, even though a basic variable ($s_1$) was available, because $s_1$ would have been negative in the corresponding basic solution. However, if we are prepared to drop the requirement that tableaux be primal feasible throughout Phase I, we can avoid the introduction of superfluous artificial variables. Since Phase II requires primal feasibility, we must ensure that negative values in the resource column eventually disappear. This can be realised by including in $z^{\mathrm{I}}$ the sum of the basic variables with negative values. Thus, if we put $\delta_i = -1$ when $x_{B_i}$ is an artificial variable, $\delta_i = 1$ when $\alpha_{io} < 0$ (which means $x_{B_i}$ will be a slack variable) and $\delta_i = 0$ otherwise, then Phase I involves maximising $z^{\mathrm{I}}$, which we will call the **infeasibility**, where

$$z^{\mathrm{I}} = \sum_i \delta_i x_{B_i}$$

$$= \sum_i \delta_i (\alpha_{io} - \sum_j \alpha_{ij} x_{N_j})$$

so that the entries in the objective row are $\alpha_{0j}$, where

$$\alpha_{0j} = \sum_i \delta_i \alpha_{ij} \qquad \text{for all } j. \tag{3.2}$$

In words, the objective row is the sum of the rows with basic slack variables and negative RHSs minus the sum of the rows with basic artificial variables. Note that, if the resource column is non-negative, $z^{\mathrm{I}}$ contains only artificial variables, whereas if $\alpha_{io} < 0$ for any $i \geqslant 0$ then $z^{\mathrm{I}} < 0$ and the corresponding tableau cannot be optimal for maximising $z^{\mathrm{I}}$.

For example, the initial Phase I tableau for P5 is P5/T1. (Since equality constraints can always be entered in the tableau with a non-negative RHS we have multiplied the third constraint by $-1$.)

P5: maximise $x_1 - 4x_2 - 5x_3$

subject to

$$x_1 - 2x_2 + x_3 \leqslant -1$$
$$2x_1 - x_2 + 2x_3 \geqslant 7$$
$$x_1 - 3x_2 - 2x_3 = -6$$
$$2x_1 - x_2 + x_3 \geqslant 7$$
$$x_1, x_2, x_3 \geqslant 0.$$

| P5 | $x_1$ | $x_2$ | $x_3$ | T1 |
|---|---|---|---|---|
| $s_1$ | 1 | −2 | 1 | −1 |
| $s_2$ | −2 | 1 | −2 | −7 |
| $a_3$ | −1 | 3 | (2) | 6 |
| $s_4$ | −2 | 1 | −1 | −7 |
| $z^{\mathrm{I}}$ | −2 | −3 | −4 | −21 |

The infeasibility $z^{\mathrm{I}}$ is a function of the basic variables having negative RHSs and these will change from tableau to tableau. Consequently, $z^{\mathrm{I}}$ is tableau-dependent and the objective row will not necessarily be correctly obtained by pivoting. Instead, (3.2) should be used in each tableau of Phase I.

The lack of primal feasibility in Phase I means that our criterion for selecting the pivot row will need revision. The approach we shall adopt is to require that no currently non-negative basic variable is allowed to become negative. In other words, we must maintain primal feasibility in those rows which are already primal feasible. This means that, if $\alpha_{io} \geqslant 0$, we require $\alpha'_{io} \geqslant 0$, and the argument of Section 2.4 applies to the primal feasible rows. Thus, we will pivot in row $r$, where $\alpha_{rk} > 0, \alpha_{ro} \geqslant 0$ and

$$\frac{\alpha_{ro}}{\alpha_{rk}} = \min_{i \geqslant 1} \left\{ \frac{\alpha_{io}}{\alpha_{ik}} \,\middle|\, \alpha_{ik} > 0,\ \alpha_{io} \geqslant 0 \right\}. \tag{3.3}$$

The pivot column is chosen as in Section 3.1 (most negative $\alpha_{0j}$) and these two rules lead to the pivot indicated in P5/T1. Pivoting gives P5/T2.

| P5 | $x_1$ | $x_2$ | $a_3$ * | T2 |
|---|---|---|---|---|
| $s_1$ | $\frac{3}{2}$ | $\frac{7}{2}$ | $-\frac{1}{2}$ | −4 |
| $s_2$ | −3 | 4 | 1 | −1 |
| $x_3$ | $-\frac{1}{2}$ | $\frac{3}{2}$ | $\frac{1}{2}$ | 3 |
| $s_4$ | ($-\frac{5}{2}$) | $\frac{5}{2}$ | $\frac{1}{2}$ | −4 |
| $z^{\mathrm{I}}$ | −4 | 3 | | −9 |

In P5/T2 we wish to pivot in the $x_1$-column, but there is no pivot according to the rule (3.3). More generally, whenever $\alpha_{ik} \leqslant 0$ for all $i$ satisfying $\alpha_{io} \geqslant 0$, this rule will not specify a pivot and a modified rule is needed. The modification relies on the observation that, if we pivot in any row $r$ for which $\alpha_{rk} < 0$ and $\alpha_{ro} < 0$, then $\alpha'_{ro} > 0$. In addition, if $\alpha_{io} \geqslant 0$, then

$$\alpha'_{io} = \alpha_{io} - \frac{\alpha_{ik}\,\alpha_{ro}}{\alpha_{rk}} \geqslant \alpha_{io} \geqslant 0,$$

since $\alpha_{ik} \leqslant 0$, by assumption. In other words, all non-negative entries in the resource column remain non-negative and at least one negative entry becomes positive. When more than one row satisfies this requirement we will choose the row that leads to the greatest increase in $z^{\mathrm{I}}$. This increase is $-\alpha_{0k}\,\alpha_{ro}/\alpha_{rk}$ and, since $\alpha_{0k} < 0$, it is maximised by choosing $r$ to satisfy $\alpha_{rk} < 0, \alpha_{ro} < 0$ and

$$\frac{\alpha_{ro}}{\alpha_{rk}} = \max_{i \geqslant 1} \left\{ \frac{\alpha_{io}}{\alpha_{ik}} \,\middle|\, \alpha_{ik} < 0, \alpha_{io} < 0 \right\}. \tag{3.4}$$

This rule also maximises the number of negative RHSs which become non-negative (Exercise 6). We will call it the Phase I PRS rule.

Applying (3.4) to P5/T2 gives the indicated pivot and leads to P5/T3, in which (3.3) gives the pivot shown (we could also have chosen the $x_3$-row). Pivoting results in P5/T4, in which, incidentally, $s_1$ switches from negative to positive in the basic solution. Consequently P5/T4 is primal feasible (although degenerate) and Phase I is completed.

| P5 | $s_4$ | $x_2$ | T3 |
|---|---|---|---|
| $s_1$ | $\frac{3}{5}$ | $-2$ | $-6\frac{2}{5}$ |
| $s_2$ | $-\frac{6}{5}$ | (1) | $3\frac{4}{5}$ |
| $x_3$ | $-\frac{1}{5}$ | 1 | $3\frac{4}{5}$ |
| $x_1$ | $-\frac{2}{5}$ | $-1$ | $1\frac{3}{5}$ |
| $z^{\mathrm{I}}$ | $\frac{3}{5}$ | $-2$ | $-6\frac{2}{5}$ |

| P5 | $s_4$ | $x_3$ | T4 |
|---|---|---|---|
| $s_1$ | $\frac{1}{5}$ | 2 | $1\frac{1}{5}$ |
| $s_2$ | $-1$ | $-1$ | 0 |
| $x_2$ | $-\frac{1}{5}$ | 1 | $3\frac{4}{5}$ |
| $x_1$ | $-\frac{3}{5}$ | 1 | $5\frac{2}{5}$ |
| $z$ | $\frac{1}{5}$ | 2 | $-9\frac{4}{5}$ |

In general, in Phase I, (3.3) is used whenever it specifies a pivot row and (3.4) is used otherwise. Alternatively, (3.4) can be used whenever the RHS of (3.4) exceeds the RHS of (3.3).

Whenever rule (3.4) is used it must specify a pivot row, for there will always be at least one $i$ for which $\alpha_{ik} < 0$ and $\alpha_{io} < 0$, if (3.3) fails. To see this note that, in an obvious notation,

$$\alpha_{0k} = \sum_{i \geqslant 1} \delta_i \alpha_{ik} = \sum_{\delta_i = 1} \alpha_{ik} - \sum_{\delta_i = -1} \alpha_{ik}.$$

Now $\alpha_{0k} < 0$, and the failure of (3.3) means that $\alpha_{ik} \leqslant 0$ whenever $\alpha_{io} \geqslant 0$ and, in particular, whenever $\delta_i = -1$, so that $\sum_{\delta_i = 1} \alpha_{ik} < 0$. But $\delta_i = 1$ whenever $\alpha_{io} < 0$ and so at least one $\alpha_{ik}$ must be negative in this sum.

The application of (3.3) or (3.4) continues until a primal feasible tableau is reached in which all artificial variables have been removed. Then Phase II is commenced by calculating the objective row for the original (non-artificial objective function). In problem P3, we did this by substituting for the basic variables, using the canonical equations corresponding to the final tableau of Phase I. In general, we have, for $i \geqslant 1$

$$x_{B_i} = \alpha_{io} - \sum_j \alpha_{ij} x_{N_j}. \tag{3.5}$$

Let us write $c_{B_i}$ ($c_{N_j}$) for the coefficient of $x_{B_i}$ ($x_{N_j}$) in the objective function. In problem P5, comparing P5 and P5/T4 shows that $c_{B_1} = 0, c_{B_2} = 0, c_{B_3} = -4$, $c_{B_4} = 1, c_{N_1} = 0, c_{N_2} = -5$. Notice that, since slack variables do not occur in the objective function, $c_{B_i}(c_{N_j}) = 0$ if $x_{B_i}(x_{N_j})$ is a slack variable.

In this notation, $z$ can be written, using (3.5), as

$$z = \sum_i c_{B_i} x_{B_i} + \sum_j c_{N_j} x_{N_j}$$
$$= \sum_i c_{B_i} \alpha_{io} - \sum_{i,j} c_{B_i} \alpha_{ij} x_{N_j} + \sum_j c_{N_j} x_{N_j}$$

and, comparing this expression for $z$, with the objective row equation

$$z = \alpha_{00} - \sum_j \alpha_{0j} x_{N_j},$$

we may conclude that

$$\alpha_{0j} = \sum_{i \geq 1} c_{B_i} \alpha_{ij} - c_{N_j} \qquad \text{for } j \geq 1 \tag{3.6a}$$

$$\alpha_{00} = \sum_{i \geq 1} c_{B_i} \alpha_{io}\,. \tag{3.6b}$$

In words, these formulae say that the objective row can be calculated by multiplying each row by the objective function coefficient of its basic variable, adding the results and subtracting the objective function coefficients of the non-basic variables (or zero, for the resource column). Thus, in the example, we find

$$s_4\text{-column}: \quad (-4) \times (-\tfrac{1}{5}) + 1 \times (-\tfrac{3}{5}) - 0 = \tfrac{1}{5}$$
$$x_3\text{-column}: \quad (-4) \times 1 + 1 \times 1 - (-5) = 2$$
$$\text{resource column}: (-4) \times 3\tfrac{4}{5} + 1 \times 5\tfrac{2}{5} = -9\tfrac{4}{5}$$

which have been entered in the objective row of P5/T4.

Fortuitously, P5/T4 satisfies the optimality conditions with optimal solution $(x_1, x_2, x_3) = (5\frac{2}{5}, 3\frac{4}{5}, 0)$ and $z = -9\frac{4}{5}$. Usually, we are not so lucky and Phase II requires further simplex iterations.

## 3.4 FINITENESS OF THE SIMPLEX METHOD

In this section we will investigate the question of whether the simplex method is sure to find an optimal solution of any LP. This question can be answered in the affirmative, at least for bounded, feasible problems, if (i) we can always find a pivot in any tableau which is not optimal and (ii) we will not keep on pivoting indefinitely. We will start by considering Phase II.

On point (i), a little reflection will show that the only occurrence which can prevent us finding a pivot in a Phase II tableau which is not optimal is if we are forced to pivot in a column $k$ with $\alpha_{0k} < 0$ (otherwise $k$ would not be a potential pivot column) and $\alpha_{ik} \leq 0$ for $i \geq 1$ (so that no pivot can be found). We saw, in Section 2.3, that a consequence of $\alpha_{ik} \leq 0$ for $i \geq 1$ is that, for any

$\lambda \geqslant 0$, there is a feasible solution with $x_{N_k} = \lambda$ and $x_{N_j} = 0$ for $j \neq k$. If this solution is substituted into the objective function equation (3.1), we find $z = \alpha_{00} - \alpha_{0k}\lambda$ which, because $\alpha_{0k} < 0$, shows that $z$ can be made arbitrarily large. This says that the problem is unbounded. Thus, *inter alia*, we now have a means of recognising an unbounded problem.

For example, if the objective in problem P1 had been to maximise $2x_1 - x_2$ (giving problem P6), two pivots in the simplex method lead, as the reader should confirm, to the tableau P6/T1. We are forced to seek a pivot in the $s_2$-column but none is available. Indeed, apart from the objective function row, this tableau is identical with T6 in Chapter 2 where we saw that $(x_1, x_2, s_1, s_2, s_3) = (2\frac{1}{2} + \frac{1}{2}\lambda, \frac{3}{4} + \frac{3}{4}\lambda, 0, \lambda, 9)$ is feasible for all $\lambda \geqslant 0$. We can verify then, by substitution in the objective function or directly from the tableau, that the objective function value of this solution is $z = 4\frac{1}{4} + \frac{1}{4}\lambda$. This corresponds to increasing $z$ along the unbounded edge of the feasible region in Fig. 2.1, starting at G $(\lambda = 0)$.

| P6 | $s_2$ | $s_1$ | T1 |
|---|---|---|---|
| $x_2$ | $-\frac{3}{4}$ | $\frac{1}{4}$ | $\frac{3}{4}$ |
| $x_1$ | $-\frac{1}{2}$ | $\frac{1}{2}$ | $2\frac{1}{2}$ |
| $s_3$ | 0 | 1 | 9 |
| $z$ | $-\frac{1}{4}$ | $\frac{3}{4}$ | $4\frac{1}{4}$ |

To answer point (ii) of our introductory paragraph (whether pivoting can continue indefinitely), we start by recalling that, after pivoting in row $r$ and column $k$, $\alpha_{00}' - \alpha_{00} = \alpha_{0k}\alpha_{r0}/\alpha_{rk}$ and, if column $k$ is chosen as pivot column in the simplex method, we must have $\alpha_{0k} < 0$. Since $\alpha_{rk}$ is the pivot, it must be positive and, assuming non-degeneracy, $\alpha_{r0} > 0$. Therefore, $\alpha_{00}' > \alpha_{00}$. This says that the objective function increases after each pivot and one consequence is that we can never return, after a sequence of simplex iterations, to a previously encountered tableau. Since there is only a finite number of ways we can select a basis and therefore only a finite number of distinct tableaux, we may conclude that eventually we must reach, either an optimal tableau, or a tableau from which we can deduce that the LP is unbounded.

A sequence of simplex iterations starting at and returning to the same tableau is called a **cycle** and, as we have seen, can only occur if all the tableaux involved are degenerate. If this happens, a computer program will traverse the cycle repeatedly until it is halted and never locate an optimal solution. In Section 3.6 we will see how cycling can be avoided, but in the meantime, we can put this apparently worrying possibility into perspective by stating that problems which arise in practice very rarely (if ever) cycle. Writers of computer codes for the simplex method have sufficient confidence in this statement that they do not usually include anti-cycling measures in the program. Cycling is largely a theoretical problem, although highly degenerate problems may be slow to reach an optimal tableau.

In Phase I we saw, in Section 3.3, that there is always a pivot. However, because $z^{\mathrm{I}}$ may change from tableau to tableau it is possible that the objective function value may decrease, but this can only happen when a previously negative-valued basic variable becomes positive. Thus, assuming non-degeneracy, either $z^{\mathrm{I}}$ increases or the number of negative entries in the resource column decreases (or both). Since non-negative entries in the resource column remain non-negative we can conclude that cycling cannot occur in Phase I. For a feasible LP, Phase I must terminate with $z^{\mathrm{I}} = 0$ and non-degeneracy implies that all artificial variables must be non-basic and can be dropped. Hence, Phase II can be commenced. Thus for a feasible, bounded, non-degenerate LP, the simplex method will find the optimal solution in a finite number of pivots. Furthermore, there will be an optimal BFS. In Section 3.6 we will remove the restriction of non-degeneracy.

## 3.5 ALTERNATIVE OPTIMAL SOLUTIONS

In this section we shall look at the optimality conditions for the simplex method in more detail. If $\alpha_{j0} > 0$ for all $j \geqslant 1$, then, if $x_{N_j} \geqslant 0$ for all $j$,

$$z = \alpha_{00} - \sum_j \alpha_{0j} x_{N_j} < \alpha_{00}$$

except when $x_{N_j} = 0$ for all $j \geqslant 1$. This implies that the BFS defined by the optimal tableau is uniquely optimal. In other words, an objective row of all positive elements implies unique optimality. Conversely, if $\alpha_{0k} = 0$ for some $k \geqslant 1$ in an optimal tableau and we pivot in column $k$ (assuming this to be possible), then

$$\alpha'_{00} = \alpha_{00} - \frac{\alpha_{r0}\,\alpha_{0k}}{\alpha_{rk}} = \alpha_{00},$$

so that the new tableau has the same objective function value and is, therefore, also optimal. Provided $\alpha_{r0} > 0$ (non-degeneracy implies this), the new tableau will define a new optimal BFS, thus indicating the existence of alternative optima. The tableau T1* illustrates this idea. Tableau T1* is optimal because of the non-negative objective row but the zero in the first column and the non-degeneracy indicate alternative optimal solutions. Pivoting in the first column we get the new optimal tableau T2*. Pivoting in the first column of T2* returns us to T1*. Thus there are two optimal BFSs, namely $(x_1, x_2, x_3) = (0, 3\frac{2}{3}, 4\frac{1}{3})$ and $(3\frac{1}{3}, 7, 7\frac{2}{3})$.

If there are two or more zeros in the bottom row of an optimal tableau, there will be at least two more alternative optimal BFSs, assuming non-degeneracy

| | $x_1$ | $s_3$ | $s_2$ | T1* |
|---|---|---|---|---|
| $s_1$ | 2 | $\frac{2}{3}$ | $\frac{1}{3}$ | $6\frac{2}{3}$ |
| $x_3$ | −1 | $\frac{1}{3}$ | $\frac{2}{3}$ | $4\frac{1}{3}$ |
| $x_2$ | −1 | $\frac{2}{3}$ | $\frac{1}{3}$ | $3\frac{2}{3}$ |
| $z$ | 0 | 1 | 1 | 8 |

| | $s_1$ | $s_3$ | $s_3$ | T2* |
|---|---|---|---|---|
| $x_1$ | $\frac{1}{2}$ | $\frac{1}{3}$ | $\frac{1}{6}$ | $3\frac{1}{3}$ |
| $x_3$ | $\frac{1}{2}$ | $\frac{2}{3}$ | $\frac{5}{6}$ | $7\frac{2}{3}$ |
| $x_2$ | $\frac{1}{2}$ | 1 | $\frac{1}{2}$ | 7 |
| $z$ | 0 | 1 | 1 | 8 |

and, possibly, pivoting from the new tableaux will yield further optimal BFSs. Provided some means of avoiding duplicating solutions is used, it is possible to obtain all the optimal BFSs of a problem with alternative optima (see Exercise 7). However, the effort required to complete such a task would rarely be warranted in a practical problem.

It is possible to have an optimal tableau with $\alpha_{0k} = 0$ but $\alpha_{ik} \leqslant 0$ for all $i \geqslant 0$, so that there is no pivot. The investigation and interpretation of such tableaux are left to the reader (in Exercise 8).

## 3.6 AVOIDANCE OF CYCLING

In Section 3.4 we proved that the simplex method is finite provided degeneracy does not occur. That degeneracy *can* lead to cycling is demonstrated in Exercise 9. We have also seen (in Exercise 7 of Chapter 2) that degeneracy is associated with ties in the PRS rule and in this section we will show that it is always possible to break such ties so as to avoid cycling.

We will start by considering an LP with '$\leqslant$' constraints and $b_i \geqslant 0$, so that the initial tableau has $x_{B_i} = b_i \geqslant 0$ and is therefore primal feasible. Since pivoting consists of adding multiples of one row to another, it follows that, if T is a tableau (not necessarily primal feasible) which can be reached by pivoting from the initial tableau and $\alpha_{\mathrm{T}r}$ is the element in the resource column of row $r$ of T ($\alpha_{r0}$ in T), then $\alpha_{\mathrm{T}r}$ can be written in the form $\alpha_{\mathrm{T}r} = \sum_i \mu_i b_i$, where $\mu_i$ depends on T and $r$. Furthermore, if we had $\mu_i = 0$ for all $i$, this would mean $\alpha_{\mathrm{T}r} = 0$ for any $b_1, \ldots, b_m$. But this cannot be true since, if $x_{B_r}$ is the basic variable in row $r$ of T, we can put $x_{B_r} = 1$ and set all other variables to zero. Then $b_1, \ldots, b_m$

could be chosen to make this solution feasible, giving $\alpha_{Tr} = 1$, in contradiction of $\alpha_{Tr} = 0$. Hence, at least one $\mu_i$ must be non-zero.

Now, with the given $b_1, \ldots, b_m$, we must have $\alpha_{Tr}$ positive, negative or zero. So $\sum_i \mu_i b_i > 0$, or $< 0$, or zero. We could write down a similar inequality or equation for every basic variable in every tableau. Then we could increase $b_1$ by a small amount so that all '>' ('<') inequalities remain '>' ('<') inequalities and all equalities with $\mu_1 \neq 0$ become inequalities. We can add the newly created inequalities to our existing set of inequalities and then increase $b_2$, continuing the procedure with $b_3, \ldots, b_n$. Eventually, because at least one $\mu_i$ is non-zero, every equality must disappear in this process.

This argument shows that it is possible to adjust the RHSs of the LP so that all signed resource column entries retain their sign and all zero entries acquire a sign. In addition, we have *increased* the RHSs, so the initial tableau remains primal feasible. Therefore, we can apply Phase II of the simplex method to the adjusted LP. Any pivot satisfying the PRS rule in the adjusted problem also satisfies the PRS rule in the original problem, for otherwise applying the pivot to the original problem would give a negative resource column entry which would stay negative in the adjusted problem and therefore violate the PRS rule. The simplex method for the adjusted problem must either discover unboundedness, in which case the original problem is also unbounded since only the pivot column (the same in both problems) is involved, or an optimal tableau is reached, since the adjusted problem has no degenerate tableaux, by construction. In the latter case, if we use the pivots of the adjusted problem to solve the original problem we must reach an optimal tableau.

The argument needs slight modification when Phase I is used. We start by adjusting the RHSs as described above and then apply Phase I to the adjusted problem whilst also using the same pivots on the original problem. When the value of any basic variable changes from negative to non-negative at some pivot, in the original problem, we start the process again from the resulting tableau. (We cannot return to any preceding tableau by the argument of Section 3.4.) Starting again means increasing the resource column entries of the new tableau to avoid degeneracy and then proceeding as before. A similar adjustment is used at the start of Phase II. Note that, if a tableau is not primal feasible in the original problem, a sign-change of a basic variable value must eventually occur, since negative values in the original problem are also negative in the adjusted problem because of the absence of degeneracy. However, a sign-change in the adjusted problem implies a sign-change in the original problem. (The converse is not true and this is the reason for introducing the modification.) Since the inclusion of equality constraints does not invalidate the argument, we can conclude that the simplex method can always be made finite by a suitable choice of pivot.

This result has considerable theoretical value, as we shall see in Chapters 5 and 12. It allows us to conclude that, *if an LP has an optimal solution, it has an optimal BFS*. This generalises the result on optimal vertices described in Section 1.3. We may also deduce that *any optimal BFS is the basic solution of some optimal tableau* (having non-negative objective row). For, we can start from any tableau corresponding to the BFS and apply the method to reach an optimal tableau. The objective function value cannot increase, since the BFS is optimal, so we must be making degenerate pivots (a zero in the pivot row and resource column). Consequently the resource column does not change, so the optimal tableau corresponds to the same BFS. (This result will be used in Chapter 9.)

To carry out the adjustment procedure described above computationally is obviously impracticable. An implementable version of the method is outlined in Exercise 10. But even this is computationally expensive and rarely used in practice. The virtue of the procedure rests in its theoretical consequences, most importantly, the duality theorem of Chapter 5.

## EXERCISES

1. For each of the following LPs, find an optimal solution if one exists or show that the problem is unbounded or infeasible. Use the simplex method (in its two-phase version, when necessary).

(i) maximise $5x_1 + 6x_2$
subject to
$$2x_1 + 3x_2 \leqslant 18$$
$$2x_1 + x_2 \leqslant 12$$
$$x_1 + x_2 \leqslant 8$$
$$x_1, x_2 \geqslant 0.$$

(ii) minimise $2x_1 - 5x_2 - 3x_3$
subject to
$$x_1 - 3x_2 + 3x_3 \geqslant -2$$
$$2x_1 + x_2 - 3x_3 \geqslant -5$$
$$2x_1 - 2x_2 - x_3 \leqslant 5$$
$$x_1, x_2, x_3 \geqslant 0.$$

(iii) maximise $3x_1 + x_2 + x_3$
subject to
$$x_1 - x_2 - 4x_3 \leqslant 5$$
$$5x_1 + 2x_2 + 4x_3 \leqslant 7$$
$$x_1, x_3 \geqslant 0 \quad (x_2 \text{ free}).$$

(iv) maximise $3x_1 - x_2 + 5x_3$
subject to
$$2x_1 + 5x_2 - 8x_3 = 1$$
$$-x_1 - 2x_2 + 3x_3 = 1$$
$$x_1, x_2, x_3 \geqslant 0.$$

(v) minimise $2x_1 + 2x_2 + 5x_3$
subject to
$$10x_1 - 2x_2 + 5x_3 \geqslant 12$$
$$-3x_1 - 2x_2 - 3x_3 = -6$$
$$x_1, x_2, x_3 \geqslant 0.$$

(vi) minimise $26x_1 + 7x_2$
subject to
$$3x_1 + 2x_2 \geqslant 15$$
$$3x_1 + 5x_2 \geqslant 29$$
$$5x_1 + 2x_2 \geqslant 23$$
$$x_1, x_2 \geqslant 0.$$

Do any of these problems have alternative optimal solutions?

2. Modify the simplex method to allow for free variables. Illustrate your modifications by re-solving Exercise 1 part (iii).

3. Solve the following problem by the simplex method. In your second tableau you should have a choice of pivot rows. Try both choices and compare the successive tableaux resulting from each choice. Explain your results.

$$\begin{aligned} \text{maximise} \quad & 3x_1 + x_2 \\ \text{subject to} \quad & x_1 - x_2 \leqslant 4 \\ & x_1 - 2x_2 \leqslant 3 \\ & 2x_1 + 5x_2 \leqslant 20 \\ & 0 \leqslant x_1 \leqslant 5, \; x_2 \geqslant 0. \end{aligned}$$

4. For what range of values of $\alpha$ does the following system of equations have a solution in non-negative variables?

$$\begin{aligned} 3x_1 - 4x_2 + 3x_3 + x_4 - 3x_5 &= 1 \\ x_1 - 5x_2 + 3x_3 - \phantom{x_4 +{}} 4x_5 &= -2 \\ -7x_1 + 2x_2 + 5x_3 - 4x_4 + 3x_5 &= \alpha. \end{aligned}$$

[Hint: the question can be answered by solving two LPs.]

5. Use Phase I to find a solution of

$$\begin{aligned} 10x_1 + 3x_2 + 5x_3 - 3x_4 &\leqslant 11 \\ -4x_1 - x_2 - 2x_3 + x_4 &= -4 \\ 5x_1 + x_2 + 3x_3 - 2x_4 &= 6 \\ x_1, x_2, x_3, x_4 &\geqslant 0. \end{aligned}$$

Write out the canonical form corresponding to the final tableau of Phase I and deduce that the solution you have found is unique.

6. A certain LP has constraints

$$\begin{aligned} x_1 + x_2 - x_3 + 2x_4 &\geqslant 2 \\ -2x_1 + 2x_2 + 3x_3 + x_4 &\geqslant 2 \\ 3x_1 - x_2 + x_3 + 3x_4 &\leqslant 1 \\ x_1, x_2, x_3, x_4 &\geqslant 0 \end{aligned}$$

and a unique optimal solution in which $x_1 = x_3 = 0$ and $x_2, x_4 > 0$. Without using the simplex method or pivoting, find the optimal solution.

Find, by pivoting, the optimal tableau apart from the objective row. The optimal tableau is one of the following three possibilities. Use the tableau you have found to deduce which one and to determine whether the objective function is to be maximised or minimised.

(i) $-x_1 + x_2 + x_3 + x_4$
(ii) $x_1 + x_2 - x_3 + x_4$
(iii) $x_1 + x_2 + x_3 - x_4$.

7. Find *all* optimal BFSs of the following LP.

$$\begin{aligned} \text{maximise} \quad & 3x_1 + 3x_2 + 6x_3 \\ \text{subject to} \quad & x_1 + x_2 + 2x_3 \leqslant 8 \\ & 2x_1 + x_2 - x_3 \leqslant 13 \\ & x_1, x_2, x_3 \geqslant 0. \end{aligned}$$

8. What can you deduce from a non-degenerate, optimal tableau which has $\alpha_{0k} =$ and $\alpha_{ik} \leqslant 0$ for all $i \geqslant 0$?

9. By choosing the highest (smallest $r$) pivot whenever the PRS rule offers a choice of pivot row, show that cycling can occur when the simplex method is applied to the following.

$$\begin{aligned} \text{maximise} \quad & 2x_1 + 3x_2 - x_3 - 12x_4 \\ \text{subject to} \quad & -2x_1 - 9x_2 + x_3 + 9x_4 \leqslant 0 \\ & \tfrac{1}{3}x_1 + x_2 - \tfrac{1}{3}x_3 - 2x_4 \leqslant 0 \\ & x_1 + x_2 + x_3 + x_4 \leqslant 1 \\ & x_1, x_2, x_3, x_4 \geqslant 0. \end{aligned}$$

10. Assuming no Phase I is required, show that the adjustment procedure of Section 3.6 can be implemented by changing $b_i$ to $b_i + \epsilon^i$ for any sufficiently small $\epsilon > 0$. Apply this technique to solve Exercise 9, without explicitly choosing $\epsilon$, by recording the coefficients of $\epsilon^i$, $i = 1, \ldots, m$, in $m$ extra columns.

CHAPTER 4

# Computational Refinements

## 4.1 THE REVISED SIMPLEX METHOD – PRODUCT FORM

The essential features of the simplex method were set out in Chapter 3 and we will now describe some of the ways in which the time to find an optimal solution of a large-scale LP may be reduced. Such a reduction may be achieved either by decreasing the number of iterations required to achieve optimality or by performing pivoting more efficiently (or both). The possibility of pivoting with less effort arises because large practical problems nearly always exhibit two significant features: the number of variables greatly exceeds the number of constraints and the data is **sparse**. The latter term means that a large proportion of $a_{ij}$s are zero. Problems with 1 per cent, or even less, non-zeros, are not uncommon. By storing the data in a suitable form, spurious multiplication by zero is avoided at a small extra cost, and computation time is considerably reduced. Unfortunately, the simplex tableau of Chapter 3 gradually 'fills-in', in that the proportion of zeros decreases and the initial sparsity is eventually lost. Amongst many other advantages, the revised simplex method remedies this defect.

We will assume that the LP has been written in the form P1.

$$\text{P1:} \quad \text{maximise} \quad \sum_j c_j x_j \; (=z)$$

$$\text{subject to} \quad \sum_j a_{ij} x_j \; (\leqslant \text{ or } =) b_i \qquad \text{for all } i$$

$$x_j \geqslant 0 \qquad \text{for all } j.$$

The **data** for the problem consists of $a_{ij}$, $b_i$, $c_j$ for all $i$ and $j$ and a record of the nature of each constraint. We assume $b_i \geqslant 0$ in equality constraints.

To describe the method, we will adopt a slightly different notation from that used in Chapter 3. We will write the $i$th equation of any canonical form as

$$E_i : \sum_j \beta_{ij} X_j = \beta_{io}$$

so that $\beta_{ij}$ is the coefficient of $X_j$ in $E_i$, where $X_j$ may be a **structural** (not slack or artificial) **variable**, a slack variable, or (in Phase I) an artificial variable. The objective function will correspond to $E_0$ and the objective function variable, $z$, will also be included as one of the $X_j$. In the notation of Chapter 3, $\beta_{ij} = \alpha_{iq}$ if $X_j$ is the non-basic variable $x_{N_q}$. Otherwise, $\beta_{ij}$ is 0 or 1.

If we pivot to make $X_k$ basic in $E_r$ $(r \geqslant 1)$, we must divide $E_r$ by $\beta_{rk}$ $(\neq 0)$ and replace $E_i$ with $E_i - \beta_{ik} E_r/\beta_{rk}$. Consequently, the new coefficients $\beta'_{ij}$ are given, for all $i$ and $j$, by

$$\beta'_{rj} = \beta_{rj}\,\eta_r \quad \text{and} \quad \beta'_{ij} = \beta_{ij} + \beta_{rj}\,\eta_i \qquad \text{for } i \neq r \tag{4.1}$$

where $\eta_r = 1/\beta_{rk}$ and $\eta_i = -\beta_{ik}/\beta_{rk}$ for $i \neq r$. Hence, the pivoting calculations require only the knowledge of the **$\eta$-list**: $(\eta_0 \,|\, \eta_1, \eta_2, \ldots)$ and $r$, to transform the coefficients of $X_j$. For example, if $\eta = (\frac{2}{3} \,|\, -\frac{4}{3}, -\frac{1}{3})$, $r = 2$ and $(\beta_{0j} \,|\, \beta_{1j}, \beta_{2j})$ $= (-1 \,|\, 3, 2)$, then $(\beta'_{0j} \,|\, \beta'_{1j}, \beta'_{2j}) = (-1 + 2 \times \frac{2}{3} \,|\, 3 + 2 \times (-\frac{4}{3}), 2 \times (-\frac{1}{3}))$ $= (\frac{1}{3} \,|\, \frac{1}{3}, -\frac{2}{3})$.

If we record the $\eta$-list and pivot row number $r$ for each pivot, we can generate the coefficients of $X_j$ in any canonical form by starting with the data and repeatedly applying (4.1), using the $\eta$s and $r$s in the order in which they were generated. This process is known as a **forward transformation**. For example, if $c_j = 1$, $a_{1j} = 3$ and $a_{2j} = 2$, so that initially we have $(\beta_{0j} \,|\, \beta_{1j}, \beta_{2j}) = (-1 \,|\, 3, 2)$ and, if the first pivot results in $\eta = (\frac{2}{3} \,|\, -\frac{4}{3}, -\frac{1}{3})$, $r = 2$ and the second pivot in $\eta' = (\frac{1}{5} \,|\, -\frac{3}{5}, -\frac{2}{5})$, $r' = 1$, then one application of (4.1) gives $(\beta'_{0j} \,|\, \beta'_{1j}, \beta'_{2j}) =$ $(\frac{1}{3} \,|\, \frac{1}{3}, -\frac{2}{3})$, as we have seen, and a second application gives $(\beta''_{0j} \,|\, \beta''_{1j}, \beta''_{2j}) =$ $(\frac{2}{5} \,|\, -\frac{1}{5}, -\frac{4}{5})$.

To carry out a simplex iteration we need to calculate (i) the objective row, in order to determine the variable, $X_k$, which is to enter the basis and (ii) $\beta_{ik}$ and $\beta_{i0}$ for all $i \geqslant 1$, in order to determine the equation, $E_r$, in which $X_k$ is to be made basic. We can then use $r$ and $\beta_{ik}$ for $i \geqslant 0$ to determine the $\eta$-list which carries out the pivoting transformation. We will use (4.1) to update $\beta_{i0}$ for $i \geqslant 0$ since it is needed at each iteration, but $\beta_{ik}$ for $i \geqslant 1$ can be calculated by a forward transformation, as described above. The excess of variables over equations in most large problems means that many variables never enter the basis. Consequently, we never need to transform the corresponding columns, unlike the tableau format of Chapter 3, where all columns are transformed. However, we still need the objective row and, though this could be obtained by forward transformation for each variable, any advantage of the new method would be lost, thereby. Instead, we will describe a less costly way of obtaining the $\beta_{0j}$ for $j \geqslant 1$.

We will first write $\beta_{0j} = \sum\limits_{i \geqslant 0} e_i\,\beta_{ij}$, where $e_0 = 1$ and $e_i = 0$ for $i \geqslant 1$. Now suppose that $\bar{\beta}_{ij}$ refers to the previous canonical form, that is, $\beta_{ij}$ is obtained from $\bar{\beta}_{ij}$ using (4.1). Then,

$$\beta_{0j} = e_r \bar{\beta}_{rj} \eta_r + \sum_{i \neq r} e_i (\bar{\beta}_{ij} + \bar{\beta}_{rj} \eta_i)$$

$$= \sum_{i \geq 0} e'_i \bar{\beta}_{ij}, \tag{4.2}$$

where

$$e'_r = \sum_{i \geq 0} \eta_i e_i \quad \text{and} \quad e'_i = e_i \quad \text{for } i \neq r. \tag{4.3}$$

By repeating the argument, $\beta_{0j}$ is expressed as a similar sum to (4.2) but with $e'$ replaced by $e''$, given by applying (4.3) twice, and $\bar{\beta}$ replaced by $\bar{\bar{\beta}}$, referring to the canonical form two iterations previously. This procedure is continued until we reach $\beta_{ij}$s in (4.2) referring to the original tableau. Thus, we apply (4.3) repeatedly, using the $\eta$-lists in reverse order to that in which they were generated, and call the result $\pi_i$ for $i \geq 0$. This process is called a **backwards transformation**. For example, with the two $\eta$-lists given above, we obtain $(e'_0 \mid e'_1, e'_2) = (1 \mid \frac{1}{5} + (-\frac{3}{5}) \times 0 + (-\frac{4}{5}) \times 0, 0) = (1 \mid \frac{1}{5}, 0)$ and $(\pi_0 \mid \pi_1, \pi_2) = (1 \mid \frac{1}{5}, (\frac{2}{3}) \times 1 + (-\frac{4}{3}) \times \frac{1}{5} - \frac{1}{3} \times 0) = (1 \mid \frac{1}{5}, \frac{2}{5})$.

We have seen that (4.2) becomes

$$\beta_{0j} = \sum_{i \geq 0} \pi_i \beta^0_{ij} \tag{4.4}$$

where $\beta^0_{ij}$ refers to the original tableau. However, the backwards transformation starts with $e_0 = 1$ and, because $r = 0$ is never chosen as pivot row, this value never alters, by (4.3). Hence $\pi_0 = 1$. Thus (4.4) can be written, by noting that $\beta^0_{0j} = -c_k$ if $X_j$ is $x_k$, as

$$\beta_{0j} = \sum_{i \geq 1} \pi_i a_{ik} - c_k \qquad \text{if } X_j \text{ is the structural variable } x_k$$

$$\beta_{0j} = \pi_i \qquad \text{if } X_j \text{ is the slack variable } s_i.$$

This method of finding $\beta_{0j}$, called **pricing out** $X_j$, is applied to the non-basic variables, since $\beta_{0j} = 0$, if $X_j$ is basic. We will sometimes refer to $\beta_{0j}$ as the **price** of $X_j$. Since these formulae use the original data, the sparsity of that data can be particularly exploited in pricing.

In Phase I we minimise $z^{\mathrm{I}}$ (as in Section 3.3) so that we required $\beta_{0j} = \sum_{i \geq 1} \delta_i \beta_{ij}$, where $\delta_i$ is defined in Section 3.3 and $X_j$ is non-basic. This means that we can use (4.4) again except that we start the backwards transformation with

$$\begin{aligned} e_i &= -1 && \text{if } E_i \text{ is an equality constraint} \\ &= 1 && \text{if } E_i \text{ is an inequality constraint and } b_i < 0 \\ &= 0 && \text{otherwise} \end{aligned}$$

and price out the variables using the formulae (for non-basic variables)

$$\beta_{0j} = \sum_{i \geqslant 1} \pi_i a_{ik} \quad \text{if } X_j \text{ is } x_k \quad \text{and} \quad \beta_{0j} = \pi_i \quad \text{if } X_j \text{ is } s_i.$$

Note that $e_0 = 0$ implies $\pi_0 = 0$, so we need only apply (4.3) with the sum running over $i \geqslant 1$. As a result, $\eta_0$ is not used in Phase I. However, the forward and backward transformations in Phase II do use $\eta_0$ which then refers to the original objective function. So we will make $\eta_0$ refer to the original objective function in both phases. Thus, to initiate the forward transformation in either phase we put

$$\begin{aligned} &\beta_{ij} = a_{ik} \quad \text{for } i \geqslant 1 \quad \text{and} \quad \beta_{0j} = -c_k, && \text{if } X_j = x_k, \\ &\beta_{rj} = 1 \quad \text{and} \quad \beta_{ij} = 0 \quad \text{for } i \neq r, && \text{if } X_j = s_r. \end{aligned}$$

We will illustrate the complete procedure by solving P2.

$$\begin{aligned} \text{P2:} \quad \text{maximise} \quad & -x_1 - x_2 + x_3 - 3x_4 \qquad (=z) \\ \text{subject to} \quad & -x_1 - x_2 + x_3 - 2x_4 \leqslant -3 \\ & 2x_1 \qquad - x_3 - x_4 = 2 \\ & \qquad -x_2 - 2x_3 \qquad \leqslant -1 \\ & x_1, x_2, x_3, x_4 \geqslant 0. \end{aligned}$$

In our solution we will write $b$ for $(\beta_{00} \mid \beta_{10}, \beta_{20}, \beta_{30})$, the resource column, $e$ for $(e_0 \mid e_1, e_2, e_3)$ and $\pi$ for $(\pi_0 \mid \pi_1, \pi_2, \pi_3)$. For $(\beta_{01}, \beta_{02}, \ldots, \beta_{06})$ when $X_1, X_2, \ldots, X_6$ are, respectively, $s_1, s_3, x_1, x_2, x_3, x_4$, we will write $c^{\mathrm{I}}$ in Phase I and $c$ in Phase II. Finally, we will write $a$ for $(\beta_{0k} \mid \beta_{1k}, \beta_{2k}, \beta_{3k})$ when $X_k$ is to be made basic (the pivot column). We will also use $\xrightarrow{\text{ⓘ F}}$ (resp. $\xrightarrow{\text{ⓘ B}}$) to indicate the application of (4.1), (resp. (4.3)) using the $\eta$-list and $r$ from iteration $i$. The F and B indicate that they are part of a forward or backward transformation.

*Iteration 1*

Basis: $s_1, a_2, s_3$ (basic variables in the order of the constraints)

$b = (0 \mid -3, 2, -1), e = (0 \mid 1, -1, 1) = \pi$

$c^{\mathrm{I}} = (0, 0, -3, -2, 0, -1)$, so make $x_1$ basic using the rule of 'most negative price'.

$a = (1 \mid -1, 2, 0)$, so $r = 2$ by the Phase I PRS rule described in Section 3.3.

$\eta = (-\frac{1}{2} \mid \frac{1}{2}, \frac{1}{2}, 0)$.

*Iteration 2*

Basis: $s_1, x_1, s_3$. $b \xrightarrow{\text{① F}} (-1 \mid -2, 1, -1)$

$e = (0 \mid 1, 0, 1) \xrightarrow{\text{① B}} (0 \mid 1, \frac{1}{2}, 1) = \pi$

$c^{\text{I}} = (0, 0, 0, -2, -1\frac{1}{2}, -2\frac{1}{2})$. Make $x_4$ basic.

$(3 \mid -2, -1, 0) \xrightarrow{\text{① F}} (3\frac{1}{2} \mid -\frac{5}{2}, -\frac{1}{2}, 0) = a$

$r = 1;\quad \eta = (1\frac{2}{5} \mid -\frac{2}{5}, -\frac{1}{5}, 0)$.

*Iteration 3*

Basis: $x_4, x_1, s_3$. $b \xrightarrow{\text{② F}} (-3\frac{4}{5} \mid \frac{4}{5}, 1\frac{2}{5}, -1)$

$e = (0 \mid 1, 0, 1) \xrightarrow{\text{② B}} (0 \mid 0, 0, 1) \xrightarrow{\text{① B}} (0 \mid 0, 0, 1) = \pi$

$c^{\text{I}} = (0, 0, 0, -1, -2, 0)$. Make $x_3$ basic.

$(-1 \mid 1, -1, -2) \xrightarrow{\text{① F}} (-\frac{1}{2} \mid \frac{1}{2}, -\frac{1}{2}, -2) \xrightarrow{\text{② F}} (\frac{1}{5} \mid -\frac{1}{5}, -\frac{3}{5}, -2) = a$

$r = 3;\quad \eta = (\frac{1}{10} \mid -\frac{1}{10}, -\frac{3}{10}, -\frac{1}{2})$.

*Iteration 4*

Basis: $x_4, x_1, x_3$. $b \xrightarrow{\text{③ F}} (-3\frac{9}{10}, \frac{9}{10}, 1\frac{7}{10}, \frac{1}{2})$. Phase I completed.

$e = (1 \mid 0, 0, 0) \xrightarrow{\text{③ B}} (1 \mid 0, 0, \frac{1}{10}) \xrightarrow{\text{② B}} (1 \mid \frac{7}{5}, 0, \frac{1}{10})$

$\xrightarrow{\text{① B}} (1 \mid \frac{7}{5}, \frac{1}{5}, \frac{1}{10}) = \pi$

$c = (1\frac{2}{5}, \frac{1}{10}, 0, -\frac{1}{2}, 0, 0)$. N.B. This is the *Phase II* objective row. Make $x_2$ basic.

$(1 \mid -1, 0, -1) \xrightarrow{\text{① F}} (1 \mid -1, 0, -1) \xrightarrow{\text{② F}} (-\frac{2}{5} \mid \frac{2}{5}, \frac{1}{5}, -1)$

$\xrightarrow{\text{③ F}} (-\frac{1}{2} \mid \frac{1}{2}, \frac{1}{2}, \frac{1}{2}) = a$

$r = 3;\quad \eta = (1 \mid -1, -1, 2)$.

*Iteration 5*

Basis: $x_4, x_1, x_2.\ b \xrightarrow{\text{④ F}} (-3\frac{2}{5} \mid \frac{2}{5}, 1\frac{1}{5}, 1)$

$$e = (1 \mid 0, 0, 0) \xrightarrow{\text{④ B}} \dots \xrightarrow{\text{① B}} (1 \mid \tfrac{7}{5}, \tfrac{1}{5}, -\tfrac{2}{5}) = \pi$$

$c = (1\frac{2}{5}, -\frac{2}{5}, 0, 0, 1, 0)$. Make $s_3$ basic.

$$(0 \mid 0, 0, 1) \xrightarrow{\text{① F}} \dots \xrightarrow{\text{④ F}} (-\tfrac{2}{5} \mid \tfrac{2}{5}, \tfrac{1}{5}, -1) = a$$

$r = 1;\quad \eta = (1 \mid \frac{5}{2}, -\frac{1}{2}, \frac{5}{2})$.

*Iteration 6*

Basis: $s_3, x_1, x_2.\ b \xrightarrow{\text{⑤ F}} (-3 \mid 1, 1, 2)$

$$e = (1 \mid 0, 0, 0) \xrightarrow{\text{⑤ B}} \dots \xrightarrow{\text{① B}} (1 \mid 1, 0, 0) = \pi$$

$c = (1, 0, 0, 0, 0, 1)$. We have achieved an optimal solution

$(x_1, x_2, x_3, x_4) = (1, 2, 0, 0)$.

This example is designed purely to demonstrate the method and not to display the computational advantages, which only become apparent for large, sparse problems. For such problems, the decrease in computation time as compared with the tableau format of Chapter 3 is dramatic. In addition, handling the data within a computer is more convenient in this method. (The full tableau of a large problem is too big to be contained in core storage, and back-up storage has to be used, entailing extra time whenever data is transferred from the back-up storage.) For this reason, all computer codes designed to handle large problems use some version of this **Product-Form Revised Simplex** (PFRS) method.

## 4.2 REINVERSION

The solution of P2 by the PFRS method illustrates an undesirable feature of the method. As the number of iterations increases so does the time required to perform forward and backward transformations, because an $\eta$-list is created at each iteration. For example, five applications of (4.3) were required to perform the backward transformation at iteration 6. However, only structural variables $x_1$ and $x_2$ are basic and so the corresponding tableau could have been achieved with only two pivots. Consequently the number of $\eta$-lists could be reduced to two. The two $\eta$-lists can be determined by writing down the coefficients of the basic structural variables and choosing pivots to force $s_1$ and $a_2$ (non-basic variables at iteration 6) out of the basis. The calculations are set out as follows.

| R1 | $x_1$ | $x_2$ |
|---|---|---|
| $s_1$ | $-1$ | $-1$ |
| $a_2$ | (2) | 0 |
| $s_3$ | 0 | $-1$ |
| $z$ | 1 | 1 |

| R2 | $x_2$ |
|---|---|
| $s_1$ | ($-1$) |
| $x_1$ | 0 |
| $s_3$ | $-1$ |
| $z$ | 1 |

(R1) $r = 2;\ \eta = (-\frac{1}{2} \mid \frac{1}{2}, \frac{1}{2}, 0)$

(R2) $r = 1;\ \eta = (1 \mid -1, 0, -1)$

Basis: $x_2, x_1, s_3$

Tableau R1 displays the coefficients of $x_1$ and $x_2$, the basic structural variables. We can now choose any non-zero pivot in rows 1 and 2 (since only $s_3$ is basic at iteration 6) and either column. We choose the $x_1$-column and $r = 2$ which gives the $\eta$-list (R1). The remaining column is then transformed according to (4.1), using this $\eta$-list. This gives tableau R2, in which no choice of pivot is available; we must make $s_1$ non-basic. The second $\eta$-list (R2), is the result. Note that the order of basic variables must be recorded since it is different from that of iteration 6.

The generalisation of this procedure is clear. After initially writing down the coefficients of the basic structural variables, at each stage we select a column and a suitable row (one containing a variable to be made non-basic) which contains a non-zero pivot. Then $r$ and the resulting $\eta$-list are recorded and the chosen column deleted. The remaining columns are transformed and the process repeated until all columns are dealt with. The only detail to be filled in is how to choose the pivot at each stage. We have already stressed the importance of sparsity and we would like to choose those pivots which lead to a maximum total number of zeros in the resulting $\eta$-lists. To solve this problem completely for a large LP would be very time consuming, but a simple approach which is quite effective is to choose a column with a maximum number of zeros and then, from those rows having a non-zero entry in the chosen column, select the row with the greatest number of zeros. The choice in tableau R1 is consistent with this rule and the reader should confirm that any other feasible choice would result in one less zero in the $\eta$-lists. The procedure relies on the fact that there will always be at least one non-zero in one of the suitable rows in each of the remaining columns. (The proof of this fact is left to Exercise 3.) It is called **reinversion**.

As well as reducing the number of $\eta$-lists, reinversion can help to control arithmetical errors which build up in computer calculations – the inevitable result of doing arithmetic to a finite number of decimal (or binary) places. The reduction in the number of $\eta$-lists and increase in sparsity contribute to improved accuracy and we can also attempt to avoid small pivots, which lead to large errors, in the reinversion procedure (possibly at some cost in sparsity). For this reason, we should recalculate the resource column either by a forward transformation of the RHSs after reinversion or by including an additional resource column during reinversion. The reader should check that the forward transformation of the RHS of P2, using $\eta$-lists (R1) and (R2), agrees with $b$ of iteration 6 in Section 4.1.

If large errors have occurred, it is possible that, after reinversion, $\beta_{io} < 0$ for some $i \geqslant 1$. It will then be necessary to perform some Phase I iterations to restore primal feasibility. Indeed, this suggests a method for starting large LPs, namely choosing a basic variable for each constraint and using the reinversion procedure to generate the corresponding $\eta$-lists. Then Phase I iterations are used to restore primal feasibility, followed by Phase II. The closer the chosen basis is to the optimal basis, the fewer iterations we may expect to need to reach optimality. Since problems are not solved in isolation, most modellers can suggest, after consideration of the practical problem or experience gained from previous versions of the model, a good or even nearly optimal basis and considerable savings in computation time can be made.

## 4.3 REFINED PIVOT COLUMN SELECTION

The requirement that the objective function increases after each pivot in a non-degenerate tableau imposes the restriction that $\beta_{0k}$ must be negative if $X_k$ is to enter the basis. In Chapter 3, we argued that a unit increase in $X_k$ increased the objective function by $-\beta_{0k}$ and so we chose the largest $-\beta_{0j}$. However, a unit change in $X_k$ also changes the $i$th basis variable by $-\beta_{ik}$ and we could measure the total change in all variables as

$$\delta_k = (1 + \sum_{i \geqslant 1} \beta_{ik}^2)^{1/2}.$$

If we choose the variable which gives the greatest increase in $z$ per unit change in all variables, we are led to choosing the most negative $\beta_{0j}/\delta_j$. We can view $1/\delta_j$ as a weighting factor to be applied to elements of the objective row. The use of the Euclidean form above for $\delta_j$ and, indeed, any other weighting factor stand merely as suggestions until computational experiment has proved that they are worth while. It turns out that the use of $1/\delta_j$ as a weighting factor can lead to a dramatic reduction in the number of iterations required to obtain an optimal solution (by more than 50 per cent in many cases).

When using the tableaux in Chapter 3, the calculation of $\delta_j$ is straightforward. However, in the PFRS method the $\beta_{ij}$ required are not available directly and to calculate them by a forward transformation would be prohibitively costly in computation time and nullify the advantage of fewer iterations. Nevertheless, we shall see that, at the cost of calculating the $\delta_j^2$ initially and performing two extra backwards transformations, we can calculate the $\delta_j^2$ at any iteration. We will derive a rule for updating the $\delta_j^2$ after each pivot.

For $i \neq r$, it follows from (4.1) that

$$\beta_{ij} = \beta'_{ij} + \beta'_{rj}\,\beta_{ik}$$

and, therefore, squaring and summing over $i \geqslant 1, i \neq r$ gives

$$\sum_{i \neq r} \beta_{ij}^2 = \sum_{i \neq r} (\beta'_{ij})^2 + (\beta'_{rj})^2 \sum_{i \neq r} \beta_{ik}^2 + 2\beta'_{rj} \sum_{i \neq r} \beta'_{ij} \beta_{ik},$$

so

$$\delta_j^2 - \beta_{rj}^2 - 1 = (\delta_j')^2 - (\beta_{rj}^2)^2 - 1 + (\beta'_{rj})^2 (\delta_k^2 - \beta_{rk}^2 - 1)$$
$$+ 2\beta'_{rj} (- \sum_{i \geqslant 1} \beta'_{ij} \eta_i \beta_{rk} + \beta'_{rj} \eta_r \beta_{rk})$$

and, noting that $\beta_{rj} = \beta'_{rj} \beta_{rk}$ and $\eta_r \beta_{rk} = 1$, this can be simplied to

$$(\delta_j')^2 = \delta_j^2 + 2\beta'_{rj} \beta_{rk} S_j - (\beta'_{rj})^2 \delta_k^2 \tag{4.5}$$

where $S_j = \sum_{i \geqslant 1} \eta_i \beta'_{ij}$. Now $\delta_j^2$ and $\delta_k^2$ are already known and $\beta_{rk}$ is calculated when the $\eta$-list is computed. The remaining unknowns are $\beta'_{rj}$ and $S_j$. These are both of the form $\sum_{i \geqslant 1} e_i \beta'_{ij}$ where, for $\beta'_{rj}$, we put (i) $e_r = 1$ and $e_i = 0$ for $i \geqslant 1$, $i \neq r$ and, for $S_j$, we put (ii) $e_i = \eta_i$ for $i \geqslant 1$. By the argument of Section 4.1, if we apply the backward transformation using (4.3) to obtain $\phi_i$ in case (i) and $\psi_i$ in case (ii) (in place of the $\pi_i$ of Section 4.1), then

$$\beta'_{rj} = \sum_{i \geqslant 1} \phi_i a_{ik}, \quad S_j = \sum_{i \geqslant 1} \psi_i a_{ik}, \quad \text{if } X_j \text{ is } x_k, \tag{4.6a}$$

and

$$\beta'_{rj} = \phi_i, \quad S_j = \psi_i, \quad \text{if } X_j \text{ is } s_i. \tag{4.6b}$$

Formulae (4.5) and (4.6) are used for those non-basic variables which remain non-basic. If $X_J$ is the variable which has just left the basis then

$$(\delta'_J)^2 = 1 + \sum_{i \geqslant 1} \eta_i^2,$$

since $\eta_i = \beta'_{iJ}$.

We will illustrate the method by reconsidering the solution of P2 given in Section 4.1. If we let $X_1$, $X_2$, $X_3$, $X_4$ be $x_1$, $x_2$, $x_3$, $x_4$ (we will not need weighting factors for other variables) we find that in iteration 1, $\delta_1^2 = \delta_4^2 = 6$, $\delta_2^2 = 3$ and $\delta_3^2 = 7$. Hence, $\beta_{01}/\delta_1 = -1.22$, $\beta_{02}/\delta_2 = -1.15$, $\beta_{03}/\delta_3 = -0.38$, $\beta_{04}/\delta_4 = -0.41$ and so $X_1$ (i.e. $x_1$) can enter the basis, as in the previous solution. Then, we find, by backwards transformation, $(\phi_1, \phi_2, \phi_3) = (0, \frac{1}{2}, 0)$ and $(\psi_1, \psi_2, \psi_3) = (\frac{1}{2}, \frac{1}{2}, 0)$ giving $(\beta'_{r2}, \beta'_{r3}, \beta'_{r4}) = (0, -\frac{1}{2}, -\frac{1}{2})$ and $(S_2, S_3, S_4) = (-\frac{1}{2}, 0, -\frac{3}{2})$ by (4.6a) and (4.6b). Since $\delta_k^2 = 6$ and $\beta_{rk} = 2$, we deduce from (4.5) that, at iteration 2, $(\delta_2^2, \delta_3^2, \delta_4^2) = (3, 5\frac{1}{2}, 7\frac{1}{2})$. There is no need to calculate $\delta_1^2$ (because $X_1$ is basic) or $\delta_j^2$ for the new non-basic variable (because it is artificial and therefore never allowed to re-enter the basis). At iteration 2, we now have $\beta_{02}/\delta_2 = -1.15$, $\beta_{03}/\delta_3 = -0.64$ and $\beta_{04}/\delta_4 = -0.91$ so we would make $x_2$ basic (as opposed to $x_4$ in the previous solution). The reader should complete the solution to verify that the $\eta$-list at iteration 2 concludes Phase I and gives a solution which is optimal for Phase II. No further iterations are needed. That Phase II is not required is a fluke, as no account of the true objective function is taken in Phase I, but, in general, column weighting factors have proved their worth. A similar scheme to the one outlined requires only one additional backwards transformation but only a rough approximation to $\delta_j$ is calculated.

Other tactics for pivot column selection include **multiple pricing**, in which several columns, with, say, large $-\beta_{0j}$, are chosen and subjected to forward transformation. These columns can be treated as a tableau and optimised as in Chapter 3, thereby avoiding backwards transformations. The $\eta$-lists generated are also recorded. When an optimum is reached we return to the main problem and repeat the process. Another possibility is **partial pricing** in which only a subset of the non-basic variables is priced out. The subset chosen may depend on the structure of the problem or, for example, may be the next $K$ variables after the last one considered. Notice, however, that all non-basic variables must be priced out to detect optimality.

## 4.4 REFINED PIVOT ROW SELECTION

The anti-cycling technique described in Section 3.6 is a valuable theoretical device but is impracticable for large problems. In this section we will outline a practical pivot row selection procedure.

In view of the importance of preventing large inaccuracies from building up, it is sensible to avoid very small pivots since they could magnify otherwise small errors. Thus, a tolerance $\tau$ is chosen on the basis of experience with programs and computer. Then, once the pivot column $k$ has been selected, any $\alpha_{ik} \leqslant \tau$ is temporarily set to zero before the PRS rule is employed. The possibility of selecting $\alpha_{rk} \leqslant \tau$ if this device is not used is particularly strong in a degenerate tableau (why?).

Another observation is that the data of realistic problems is almost never specified to the level of accuracy used within a computer. Although this extra precision may be advantageous when calculating inner columns (particularly when reinverting in PFRS method) it gives a spurious impression of accuracy in the resource column since these numbers represent values of variables. At worst, this can mean that extra iterations are required and, therefore, working with less precision in the resource column can reduce the number of iterations.

A simple way of achieving this is to specify a tolerance $\delta(> 0)$ on variable values. That is, we would be prepared to accept optimal values of the variables within $\delta$ of their true value. The larger $\delta$ is, the more iterations we may expect to save but the less accurate will our answer be. For example, we might specify $\delta = 0.0005$ and then print out all answers to three places of decimals.

The requirement that after pivoting $\alpha'_{i0} \geqslant 0$ (which led to the PRS rule) is replaced by $\alpha'_{i0} \geqslant -\delta$. Assuming that $\alpha_{r0} \geqslant 0$, this requirement can be restated as

$$\frac{\alpha_{i0} + \delta}{\alpha_{ik}} \geqslant \frac{\alpha_{r0}}{\alpha_{rk}} \qquad \text{where } \alpha_{ik} > 0.$$

To achieve this, we first calculate

$$L = \min_{i \geqslant 1} \left\{ \frac{\alpha_{i0} + \delta}{\alpha_{ik}} \middle| \alpha_{ik} > 0 \text{ (or } \tau) \right\}$$

and we can then pivot in any row $r$ in which $\alpha_{r0}/\alpha_{rk} \leqslant L$. How do we choose $r$ when several rows satisfy this inequality? In the light of our earlier comments on accuracy, large pivots are generally desirable and so our rule will be to choose $r$ with $\alpha_{rk} > 0$ and (breaking ties arbitrarily)

$$\alpha_{rk} = \max_{i \geqslant 1} \{\alpha_{ik} \mid \alpha_{i0}/\alpha_{ik} \leqslant L\}.$$

If $\alpha_{r0} < 0$ and $\alpha_{rk}$ is small enough, we may find $\alpha'_{r0} < -\delta$. So, in the case $-\delta \leqslant \alpha_{r0} < 0$, we reset $\alpha_{r0}$ to zero before pivoting (so the resource column does not change).

We will now solve P4 with $\delta = 0.1$. After one pivot, we reach P4/T2.

$$\begin{aligned}
\text{P4: maximise} \quad & 3x_1 - x_2 \\
\text{subject to} \quad & 3x_1 - 5x_2 \leqslant 1 \\
& 7x_1 - 6x_2 \leqslant 11 \\
& x_1 + 3x_2 \leqslant 13 \\
& 8x_1 - 6x_2 \leqslant 14 \\
& x_1, x_2 \geqslant 0.
\end{aligned}$$

| P4 | $s_1$ | $x_2$ | T2 |
|---|---|---|---|
| $x_1$ | 0.333 | −1.667 | 0.333 |
| $s_2$ | −2.333 | 5.667 | 8.667 |
| $s_3$ | −0.333 | 4.667 | 12.667 |
| $s_4$ | −2.667 | (7.333) | 11.333 |
| $z$ | 1 | −4 | 1 |

| P4 | $s_1$ | $s_4$ | T3 |
|---|---|---|---|
| $x_1$ | −0.272 | 0.227 | 2.905 |
| $s_2$ | −0.271 | −0.773 | −0.093 |
| $s_3$ | (1.364) | −0.636 | 5.459 |
| $x_2$ | −0.364 | 0.136 | 1.545 |
| $z$ | −0.455 | 0.545 | 7.182 |

Now $8.667/5.667 = 1.529$, $12.667/4.667 = 2.714$, $11.333/7.33 = 1.545$ and $L = \min \{8.767/5.667, \ldots\} = 1.547$, so the approximate PRS rule would choose $r = 4$ (as opposed to $r = 2$ for the original rule). Pivoting as indicated gives P4/T3 and one further iteration leads to an optimal tableau with $(x_1, x_2) = (3.994, 3.002)$, which rounds to $(4, 3)$. Pivoting with $r = 2$ in P4/T2 would make $s_1$ and $s_2$ non-basic and thus the optimal tableau (which has $s_3$ and $s_4$ non-basic) is reached only after two more iterations. The approximate PRS rule can only make different choices from the original PRS rule when 'near ties' would occur in pivot row selection. This is much more common in large problems and a useful decrease in computation time can result, even when $\delta$ is several orders of magnitude smaller than 0.1.

## EXERCISES

1. Use the PFRS method to solve the following problem.

$$\begin{aligned}
\text{minimise} \quad & 3x_1 + 19x_2 - 7x_3 - 3x_4 + x_5 + 4x_6 + 6x_7 - x_8 + 2x_9 \\
\text{subject to} \quad & x_1 + 2x_2 - x_3 + 2x_4 + x_5 + x_6 - x_7 + 3x_8 + x_9 \geqslant 1 \\
& 2x_1 + 7x_3 + 4x_6 \leqslant 3 \\
& -2x_2 + x_3 + x_4 + 2x_5 - x_7 + x_8 = 1 \\
& x_1, x_2, \ldots, x_9 \geqslant 0.
\end{aligned}$$

2. Solve the following LP by (i) using the reinversion procedure to construct the $\eta$-lists required to make $x_5$, $x_7$ and $x_3$ basic in constraints 1, 2 and 3 respectively, (ii) using Phase I iterations starting with $x_5$, $x_7$ and $x_3$ basic, to obtain primal feasibility, (iii) reinverting again and (iv) using Phase II iterations to reach optimality.

$$\begin{aligned}
\text{maximise} \quad & x_1 + 2x_2 + x_3 + 6x_4 + 3x_5 + 5x_6 + 5x_7 + 2x_8 + 3x_9 + \\
& \qquad x_{10} + 2x_{11} + 2x_{12} \\
\text{subject to} \quad & -x_1 + x_3 + x_4 + x_5 + 2x_6 + x_7 + 2x_8 + x_9 + \\
& \qquad 3x_{11} + 2x_{12} = 1 \\
& x_1 + x_2 + 2x_4 + x_6 - x_7 + x_8 + x_9 + \\
& \qquad x_{10} + 2x_{11} + 3x_{12} = 1 \\
& 2x_1 - x_2 + x_3 + x_4 - x_5 + x_6 + 3x_8 + \\
& \qquad x_{10} + x_{11} + x_{12} = 1 \\
& x_1, x_2, \ldots, x_{12} \geqslant 0.
\end{aligned}$$

3. Suppose that $\alpha_{rk} = 0$ in some reduced tableau. Show that the canonical equations do not have a unique solution with $x_{B_r} = 0$ and $x_{N_j} = 0$ for $j \neq k$. Use

the fact that any basis uniquely determines a tableau to deduce that no tableau can have $x_{B_i}$ for $i \neq r$ and $x_{N_k}$ as basis. (Obviously a direct pivot is ruled out, but this result precludes any other sequence of pivots leading to this basis.) Hence show that the reinversion procedure of Section 4.2 cannot fail through the absence of a non-zero pivot.

4. Apply the pivot column selection criterion of Section 4.3 to *Phase II* of the problem of Exercise 1, i.e. use 'most negative price' in Phase I and 'most negative weighted price' in Phase II. (You will still need to update weighting factors throughout Phase I.)

5. Solve the following LP (i) using the standard simplex method and (ii) using the approximate PRS rule of Section 4.4. In both cases, work to two places of decimals and in (ii) use $\delta = 0.1$.

Make $x_1$ basic in the initial tableau

$$\begin{array}{ll} \text{maximise} & x_1 + x_2 \\ \text{subject to} & 6x_1 + x_2 \leqslant 2 \\ & 7x_1 + 2x_2 \leqslant 3 \\ & 9x_1 + 5x_2 \leqslant 6 \\ & x_1, x_2 \geqslant 0. \end{array}$$

Explain your results graphically.

CHAPTER 5

# Duality and the Dual Simplex Method

## 5.1 THE DUAL PROBLEM

Any LP with only inequality constraints can be written in the **inequality form** P.

$$\text{P:} \quad \text{maximise} \quad \sum_j c_j x_j \qquad (=z)$$

$$\text{subject to} \quad \sum_j a_{ij} x_j \leqslant b_i \qquad \text{for } i = 1, \ldots, m$$

$$x_j \geqslant 0 \qquad \text{for } j = 1, \ldots, n.$$

When the data ($c_j$, $a_{ij}$, $b_i$ for all $i$, $j$) is non-negative, P can be interpreted as a production problem in which an amount $x_j$ of the $j$th of $n$ different products is manufactured from $m$ different raw materials and $b_i$ units of the $i$th raw material are available. If $a_{ij}$ is the number of units of the $i$th raw material required to produce a unit amount of product $j$, the constraints express the limited availability of raw materials. The objective function is the total revenue generated, if product $j$ is sold at price $c_j$ per unit.

The maximum revenue obtained imputes a value, and thereby an implicit or **shadow** price, to each raw material. To determine these prices we might put ourselves in the position of an entrepreneur who wishes to purchase the raw materials. If he fixes the price of material $i$ at $y_i$, he will have to pay $\sum_i b_i y_i$. He will naturally wish to minimise this, but he must offer prices which make it worth while for the manufacturer to sell. The entrepreneur will have to pay $\sum_i a_{ij} y_i$ for the ingredients of one unit of product $j$ and the manufacturer will not sell unless this sum exceeds $c_j$. Thus, the entrepreneur's problem is DP.

$$\text{DP:} \quad \text{minimise} \quad \sum_i b_i y_i \qquad (= -w)$$

$$\text{subject to} \quad \sum_i y_i a_{ij} \geqslant c_j \qquad \text{for } j = 1, \ldots, n$$

$$y_i \geqslant 0 \qquad \text{for } i = 1, \ldots, m.$$

We will call DP the **dual problem** of P and will adopt this definition of duality even when the data is not non-negative. In this context we will sometimes refer to P as the **primal problem**. The example below has P1 as primal problem and D1 as dual problem.

$$\begin{aligned}
\text{P1: maximise} \quad & 7x_1 + 8x_2 + 5x_3 + 3x_4 \\
\text{subject to} \quad & 3x_1 + 5x_2 + 2x_3 + x_4 \leqslant 21 \\
& 8x_1 + 4x_2 + 6x_3 + 3x_4 \leqslant 17 \\
& 2x_1 + 2x_2 + x_3 + x_4 \leqslant 7 \\
& x_1, x_2, x_3, x_4 \geqslant 0.
\end{aligned}$$

$$\begin{aligned}
\text{D1: minimise} \quad & 21y_1 + 17y_2 + 7y_3 \\
\text{subject to} \quad & 3y_1 + 8y_2 + 2y_3 \geqslant 7 \\
& 5y_1 + 4y_2 + 2y_3 \geqslant 8 \\
& 2y_1 + 6y_2 + y_3 \geqslant 5 \\
& y_1 + 3y_2 + y_3 \geqslant 3 \\
& y_1, y_2, y_3 \geqslant 0.
\end{aligned}$$

Notice that each constraint in P has an associated variable in D and each variable in P has an associated constraint in D.

The dual problem D can be rewritten as $D^*$ below and its dual problem is $P^*$. But $P^*$ is just a slightly disguised version of P, showing that the dual of the dual problem is the primal problem.

$$\begin{aligned}
D^*\text{: maximise} \quad & \sum_i (-b_i) y_i \\
\text{subject to} \quad & \sum_i (-a_{ij}) y_i \leqslant -c_j \\
& y_i \geqslant 0.
\end{aligned}
\qquad\qquad
\begin{aligned}
P^*\text{: minimise} \quad & \sum_j (-c_j) x_j \\
\text{subject to} \quad & \sum_j x_j (-a_{ij}) \geqslant -b_i \\
& x_j \geqslant 0.
\end{aligned}$$

Thus, duality is a symmetrical relationship and which problem of a pair we choose to focus on, by calling it the primal, is determined by the context in which the problem arises and not the mathematics of duality. Notice, however, that we always have '$\leqslant$' constraints in the 'maximisation' problem and '$\geqslant$' constraints in the 'minimisation' problem, in any dual pair.

## 5.2 THE COMPLEMENTARY SLACKNESS CONDITIONS

The problems P and D can be written, after the introduction of slack variables, in the following form

$$\text{P:} \quad \text{maximise} \quad \sum_j c_j x_j \qquad (=z)$$

$$\text{subject to} \quad \sum_j a_{ij} x_j + s_i = b_i \quad \text{all } i$$

$$x_j, s_i \geqslant 0 \quad \text{all } i, j.$$

$$\text{D:} \quad \text{minimise} \quad \sum_i b_i y_i \qquad (=-w)$$

$$\text{subject to} \quad \sum_i a_{ij} y_i - t_j = c_j \quad \text{all } j$$

$$y_i, t_j \geqslant 0 \quad \text{all } i, j.$$

From this, we see that if $z$ and $-w$ are objective function values corresponding to feasible solutions in their respective problems we must have

$$\begin{aligned}(-w) - z &= \sum_i b_i y_i - \sum_j c_j x_j \\ &= \sum_i \left( \sum_j a_{ij} x_j + s_i \right) y_i - \sum_j \left( \sum_i a_{ij} y_i - t_j \right) x_j \\ &= \sum_i s_i y_i + \sum_j t_j x_j. \end{aligned} \tag{5.1}$$

This simple equality has surprisingly far-reaching consequences and we shall now explore some of these.

Our first observation is that, since $s_i, y_i \geqslant 0$ for all $i$ and $x_j, t_j \geqslant 0$ for all $j$, both sums are non-negative so that $-w \geqslant z$. In words, the objective function value of any feasible solution of D is at least as great as the objective function value of any feasible solution of P. Thus objective function values in P are bounded above by those in D. In particular, if P is unbounded this must mean D is infeasible. From the symmetry of the duality relationship, it follows that, if D is unbounded, P is infeasible. In general, if the primal problem is unbounded, the dual problem is infeasible. Problems P2 and D2 below illustrate this:

$$\begin{array}{llrl} \text{P2:} & \text{maximise} & 3x_1 - 2x_2 & \\ & \text{subject to} & -4x_1 - 5x_2 & \leqslant -8 \\ & & x_1 - x_2 & \leqslant 1 \\ & & x_1, x_2 & \geqslant 0. \end{array}$$

$$\begin{array}{llrl} \text{D2:} & \text{minimise} & -8y_1 + y_2 & \\ & \text{subject to} & -4y_1 + y_2 & \geqslant 3 \\ & & -5y_1 - y_2 & \geqslant -2 \\ & & y_1, y_2 & \geqslant 0. \end{array}$$

The reader should verify (graphically or using the simplex method) that P2 is unbounded and D2 infeasible.

Of more practical interest is the case when both problems are feasible and therefore bounded. If we can find feasible solutions to P and D which have objective function values of $\hat{z}$ and $-\hat{w}$ respectively, and $-\hat{w} = \hat{z}$, we may conclude that the solutions are optimal in their respective problems. For, if $z$ is the objective function value of any feasible solution of P, we know that

$$z \leqslant -\hat{w} = \hat{z},$$

which says that $\hat{z}$ is the optimal objective function value. By the symmetry of duality, $-\hat{w}$ is the optimal value in D. This gives us a criterion for optimality which is independent of the simplex method. In order to make it more readily applicable, we will use (5.1) to translate the criterion into an alternative form. We see at once that $-w = z$ is the same as

$$\sum_i s_i y_i + \sum_j t_j x_j = 0.$$

Since all the variables are non-negative, this is equivalent to

$$\begin{aligned} s_i y_i &= 0 \qquad \text{for all } i \\ t_j x_j &= 0 \qquad \text{for all } j \end{aligned}$$

and $s_i y_i = 0$, for example, can be rewritten as $s_i = 0$ or $y_i = 0$. This means that the condition $-w = z$ is equivalent to

$$x_j = 0 \quad \text{or } t_j = 0 \quad \text{for each } j$$

and

$$y_i = 0 \quad \text{or } s_i = 0 \quad \text{for each } i,$$

which we will call the **complementary slackness** (CS) conditions. The 'or' in these conditions is inclusive in that it does not preclude both $x_j = 0$ and $t_j = 0$ holding (or $y_i = s_i = 0$). We will call $x_j$ and $t_j$ a pair of **complementary variables**, similarly for $y_i$ and $s_i$. Thus each complementary pair contains one variable from the primal and one from the dual and it also contains one slack and one structural variable. We can state the CS conditions concisely as the requirement that at least one variable in each complementary pair is zero.

Our discussion so far has shown that, if the CS conditions are satisfied by feasible solutions of the primal and dual problems, then these solutions are optimal. If we have a feasible solution to an LP which we wish to test for optimality, we must attempt to find a feasible solution of the dual problem satisfying the CS conditions. Note that feasibility in both problems is equivalent to all variables, $x_j$, $s_i$, $y_i$, $t_j$, being non-negative. As an example we will test $(x_1, x_2, x_3, x_4) = (0, 3\frac{1}{8}, \frac{3}{4}, 0)$ for optimality in P1. The slack variables then take the values $s_1 = 3\frac{7}{8}$, $s_2 = 0$, $s_3 = 0$. The non-negativity of $x_j$ and $s_i$ shows that the solution is feasible and, if the CS conditions are to be satisfied, we must have $t_2 = t_3 = 0$ (because $x_2, x_3 > 0$) and $y_1 = 0$ (because $s_1 > 0$). Since $t_2 = t_3 = 0$ means that the second and third constraints of D1 are satisfied as equalities and $y_1$ vanishes, $y_2$ and $y_3$ must be a solution of

$$4y_2 + 2y_3 = 8$$
$$6y_2 + \ y_3 = 5,$$

so $y_2 = \frac{1}{4}, y_3 = 3\frac{1}{2}$ and it follows, by substitution in the constraints of D1, that $t_1, t_4 \geqslant 0$ (in fact, $t_1 = 2, t_4 = 1\frac{1}{4}$). Hence the solution $(y_1, y_2, y_3) = (0, \frac{1}{4}, 3\frac{1}{2})$ is feasible in D1 and the CS conditions are satisfied, so our original solution to P1 is optimal.

As an incidental feature of our application of the CS conditions, we have obtained an optimal solution of D1. As we argued above, this solution gives the shadow prices of the raw materials imputed by the production problem. The initially rather startling fact that the first raw material has a price of zero, arises from the fact that we do not use all of this material, i.e. $s_1 > 0$. Consequently, a small reduction (or increase) in the stock of this material does not change our sales revenue and therefore has no marginal value, which is what a shadow price measures. Such an input is sometimes described as a **free good** and is yet another manifestation of complementary slackness. The further development of this interpretation is of great concern in economics, and we will return to it briefly in Section 5.5. Yet another related interpretation of duality arises in the context of RHS perturbations. This will be described implicitly in the following chapter on sensitivity analysis.

For the CS conditions to be really useful, we would expect that, if the CS conditions are not satisfied, the trial solution is not optimal. This result is true, though more profound than may appear at first sight. We will prove the result in Section 5.4 but, as a preliminary, we will explore the interrelationship between duality and the simplex method.

## 5.3 THE DUAL SIMPLEX METHOD

In this section, we will examine what happens to the tableaux corresponding to the dual problem when the simplex method is applied to the primal problem. This examination will lead to a new computational procedure, the **dual simplex method**, for solving LPs and indirectly provide the means for proving the duality theorem in Section 5.4.

To illustrate the ideas involved we shall look at the problem P3.

$$\begin{aligned} \text{P3:} \quad \text{maximise} \quad & 2x_1 + x_2 \qquad (=z) \\ \text{subject to} \quad & 3x_1 + x_2 \leqslant 12 \\ & x_1 - 2x_2 \leqslant 2 \\ & -x_1 + 3x_2 \leqslant 9 \\ & x_1, x_2 \geqslant 0 \end{aligned}$$

| P3 | $x_1$ | $x_2$ | T1 |
|---|---|---|---|
| $s_1$ | 3 | 1 | 12 |
| $s_2$ | (1) | −2 | 2 |
| $s_3$ | −1 | 3 | 9 |
| $z$ | −2 | −1 | 0 |

which has as its initial tableau P3/T1. The dual problem is D3.

$$\begin{aligned} \text{D3:} \quad & \text{minimise} & 12y_1 + 2y_2 + 9y_3 & \qquad (=-w) \\ & \text{subject to} & 3y_1 + y_2 - y_3 \geqslant 2 & \\ & & y_1 - 2y_2 + 3y_3 \geqslant 1 & \\ & & y_1, y_2, y_3 \geqslant 0. & \end{aligned}$$

D3 can be written as D3′.

$$\begin{aligned} \text{D3}': \quad & \text{maximise} \quad w = & -12y_1 - 2y_2 - 9y_3 & \\ & \text{subject to} & -3y_1 - y_2 + y_3 + t_1 &= -2 \\ & & -y_1 + 2y_2 - 3y_3 + t_2 &= -1 \\ & & y_1, y_2, y_3, t_1, t_2 \geqslant 0 & \end{aligned}$$

and the tableau corresponding to this initial, basic, but not feasible, solution is D3/T1.

| D3 | $y_1$ | $y_2$ | $y_3$ | T1 |
|---|---|---|---|---|
| $t_1$ | −3 | (−1) | 1 | −2 |
| $t_2$ | −1 | 2 | −3 | −1 |
| $w$ | 12 | 2 | 9 | 0 |

The relationship between P3/T1 and D3/T1 is interesting. Informally, we may summarise it by saying that rows and columns are interchanged, variables replaced by their complementary variables and entries, other than those in the resource column and objective row, are multiplied by $-1$. To make this more precise we will apply a superscript P or D to indicate that the variable superscripted belongs to the primal or dual problem respectively. Then we can see that to transform P3/T1 to D3/T1 involves the following replacement of variables

$$\begin{aligned} s_i \; &<\!-\!> \; y_i & \qquad i \geqslant 1 \\ x_j \; &<\!-\!> \; t_j & \qquad j \geqslant 1 \\ z \; &<\!-\!> \; -w & \end{aligned}$$

where the double-headed arrow means 'replace the item on the left by the item on the right' and is double-headed to indicate that the replacements work in reverse to obtain P3/T1 from D3/T1. The values to be inserted in D3/T1 are given by the following replacement rules.

$$\begin{aligned}
\alpha_{ij}^{D} &= -\alpha_{ji}^{P} && i, j \geqslant 1 \\
\alpha_{i0}^{D} &= \alpha_{0i}^{P} && i \geqslant 1 \\
\alpha_{0j}^{D} &= \alpha_{j0}^{P} && j \geqslant 1 \\
\alpha_{00}^{D} &= -\alpha_{00}^{P}.
\end{aligned}$$

The negative sign in the rule for $\alpha_{00}^{D}$ cannot be inferred from the tableaux above but it will be justified in the subsequent development. The rules also apply if we wish to obtain P3/T1 from D3/T1 and the symmetry between dual problems observed in the last chapter is again manifest, showing that our labelling of one problem as 'primal' and the other 'dual' is artificial. In future we shall refer to any pair of tableaux, in which each can be obtained from the other by the use of these replacement rules, as **dual tableaux.**

That P3/T1 and D3/T1 are dual tableaux in this example is not a coincidence. For, if we take the standard dual pair of LPs, P and D of Section 5.1., then we have initial tableaux in which

$$\begin{aligned}
\alpha_{ij}^{P} &= a_{ij}, & \alpha_{ji}^{D} &= -a_{ij} && i, j \geqslant 1 \\
\alpha_{i0}^{P} &= b_i = & \alpha_{0i}^{D} & && i \geqslant 1 \\
\alpha_{0j}^{P} &= -c_j = & \alpha_{j0}^{D} & && j \geqslant 1 \\
\alpha_{00}^{P} &= 0 = & \alpha_{00}^{D}, & &&
\end{aligned}$$

as the reader should verify. Thus we may state our first result by saying that *the initial tableaux corresponding to a dual pair of problems form a dual pair of tableaux*.

Let us now investigate what happens as the simplex method is applied. The first pivot in P3/T1 makes $x_1$ basic and $s_2$ non-basic and, after pivoting, we get the tableau P3/T2. The element in D3/T1 corresponding to the pivot in P3/T1 is in the $t_1$-row and $y_2$-column and pivoting on this entry, using the standard pivoting rules, gives the tableau D3/T2. We see that P3/T2 and D3/T2 again form a dual pair of tableaux.

Not surprisingly, this result is no accident. To see this, we must apply the transformation rules to a dual pair of tableaux. Let us suppose that we have such a pair and $\alpha_{rk}^{P}$, in row $r(\geqslant 1)$ and column $k(\geqslant 1)$, is the pivot in one tableau. The corresponding element in the other tableau is $\alpha_{kr}^{D} \left(= -\alpha_{rk}^{P}\right)$ in row $k$ and column $r$. We will leave the reader the straightforward task of checking that after

| P3 | $s_2$ | $x_2$ | T2 |
|---|---|---|---|
| $s_1$ | −3 | (7) | 6 |
| $x_1$ | 1 | −2 | 2 |
| $s_3$ | 1 | 1 | 11 |
| $z$ | 2 | −5 | 4 |

| D3 | $y_1$ | $t_1$ | $t_3$ | T2 |
|---|---|---|---|---|
| $y_2$ | 3 | −1 | −1 | 2 |
| $t_2$ | (−7) | 2 | −1 | −5 |
| $w$ | 6 | 2 | 11 | −4 |

pivoting in these elements in each tableau, the variables labelling rows and columns still satisfy the replacement rules above, in that they remain complementary variables. We shall denote by $\alpha_{ij}^{P'}$, $\alpha_{ij}^{D'}$, respectively, the entries in the two transformed tableaux and, combining the usual transformation rules with replacement rules for dual tableaux, we obtain

$$\alpha_{kr}^{D'} = 1/\alpha_{kr}^{D} = 1/(-\alpha_{rk}^{P}) = -\alpha_{rk}^{P'}$$

$$\alpha_{ki}^{D'} = \alpha_{ki}^{D}/\alpha_{kr}^{D} = -\alpha_{ik}^{P}/(-\alpha_{rk}^{P}) = -\alpha_{ik}^{P'} \qquad \text{for } i \geqslant 1, (i \neq r)$$

$$\alpha_{ko}^{D'} = \alpha_{ko}^{D}/\alpha_{kr}^{D} = \alpha_{ok}^{P}/(-\alpha_{rk}^{P}) = \alpha_{ok}^{P'}$$

$$\alpha_{jr}^{D'} = -\alpha_{jr}^{D}/\alpha_{kr}^{D} = \alpha_{rj}^{P}/(-\alpha_{rk}^{P}) = -\alpha_{rj}^{P'} \qquad \text{for } j \geqslant 1, (j \neq k)$$

$$\alpha_{or}^{D'} = -\alpha_{or}^{D}/\alpha_{kr}^{D} = -\alpha_{ro}^{P}/(-\alpha_{rk}^{P}) = \alpha_{ro}^{P}$$

$$\alpha_{ji}^{D'} = \alpha_{ji}^{D} - \frac{\alpha_{jr}^{D}\,\alpha_{ki}^{D}}{\alpha_{kr}^{D}} \begin{cases} = -\alpha_{ij}^{P} + \dfrac{\alpha_{ik}^{P}\,\alpha_{rj}^{P}}{\alpha_{rk}^{P}} = -\alpha_{ij}^{P'} \\ \quad \text{for } i, j \geqslant 1 \text{ or } i = j = 0\ (i, j \neq r, k) \\ = \alpha_{ij}^{P} - \dfrac{\alpha_{ik}^{P}\,\alpha_{rj}^{P}}{\alpha_{rk}^{P}} = \alpha_{ij}^{P'} \\ \quad \text{for } i \geqslant 1, j = 0 \text{ or } i = 0, j \geqslant 1\ (i, j \neq r, k). \end{cases}$$

We may conclude that the replacement rules apply to the transformed tableaux or, to summarise, *pivoting on corresponding elements in a dual pair of tableaux yields another pair of dual tableaux*.

We shall now complete our solution of the primal problem by the simplex method, setting down the dual pairs of tableaux and indicating the pivots in each. The reader is strongly urged, *for this example only*, to perform the pivoting operations for each tableau in each dual pair. This exercise will explain why the transformation operations preserve tableau duality more forcefully than any amount of perusal of the algebraic manipulations above. We get

| P3 | $s_2$ | $s_1$ | T3 |
|---|---|---|---|
| $x_2$ | $-\frac{3}{7}$ | $\frac{1}{7}$ | $\frac{6}{7}$ |
| $x_1$ | $\frac{1}{7}$ | $\frac{2}{7}$ | $3\frac{5}{7}$ |
| $s_3$ | $\frac{10}{7}$ (circled) | $-\frac{1}{7}$ | $10\frac{1}{7}$ |
| $z$ | $-\frac{1}{7}$ | $\frac{5}{7}$ | $8\frac{2}{7}$ |

| D3 | $t_2$ | $t_1$ | $y_3$ | T3 |
|---|---|---|---|---|
| $y_2$ | $\frac{3}{7}$ | $\frac{1}{7}$ | $-\frac{10}{7}$ (circled) | $-\frac{1}{7}$ |
| $y_1$ | $-\frac{1}{7}$ | $-\frac{2}{7}$ | $\frac{1}{7}$ | $\frac{5}{7}$ |
| $w$ | $\frac{6}{7}$ | $3\frac{5}{7}$ | $10\frac{1}{7}$ | $-8\frac{2}{7}$ |

| P3 | $s_3$ | $s_1$ | T4 |
|---|---|---|---|
| $x_2$ | $\frac{3}{10}$ | $\frac{1}{10}$ | $3\frac{9}{10}$ |
| $x_1$ | $-\frac{1}{10}$ | $\frac{3}{10}$ | $2\frac{7}{10}$ |
| $s_2$ | $\frac{7}{10}$ | $-\frac{1}{10}$ | $7\frac{1}{10}$ |
| $z$ | $\frac{1}{10}$ | $\frac{7}{10}$ | $9\frac{3}{10}$ |

| D3 | $t_2$ | $t_1$ | $y_2$ | T4 |
|---|---|---|---|---|
| $y_3$ | $-\frac{3}{10}$ | $\frac{1}{10}$ | $-\frac{7}{10}$ | $\frac{1}{10}$ |
| $y_1$ | $-\frac{1}{10}$ | $-\frac{3}{10}$ | $\frac{1}{10}$ | $\frac{7}{10}$ |
| $w$ | $3\frac{9}{10}$ | $2\frac{7}{10}$ | $7\frac{1}{10}$ | $-9\frac{3}{10}$ |

The tableau D3/T4 exhibits a basic solution of the dual problem which is feasible and the objective row is non-negative. In other words, it is the optimal tableau for the dual problem, the optimal dual solution is $(y_1, y_2, y_3) = (\frac{7}{10}, 0, \frac{1}{10})$ and the optimal dual objective function value is $-w = 9\frac{3}{10}$. If we describe a tableau in which $\alpha_{0j} \geqslant 0$ for $j \geqslant 1$ (non-negative objective row) as **dual feasible** then an optimal tableau is both primal and dual feasible. In Phase II of the simplex method we start from a tableau which is primal, but not usually dual, feasible and pivot, maintaining primal feasibility, until we reach a tableau which is also dual feasible and therefore optimal. The tableaux D3/T1 – T4 above run through a sequence starting with a dual feasible, but primal infeasible, tableau, maintaining dual feasibility in subsequent tableaux, and eventually reaching a tableau which is also primal feasible and therefore optimal. This is the essence of the **dual simplex method** which can be applied to any problem in which the initial tableau is dual feasible as, for example, the dual problem above.

The use of the dual simplex method avoids having recourse to Phase I of the two-phase method which would have been the only solution technique available up to now. Unfortunately, the advantage of the method is largely wasted if we have to write down a pair of tableaux at each iteration. However, the pivoting operation is standard and the only reason a pair of tableaux is required is to select the pivot. If we can translate the pivot selection rules from one member of a dual pair to the other we can dispense with dual pairs of tableaux altogether. We recall that the standard method pivots in column $k$, where

$$\alpha_{0k} = \min_{j \geqslant 1} \alpha_{0j},$$

provided $\alpha_{0k} < 0$ (otherwise the tableau is optimal) and in row $r(\geqslant 1)$, where $\alpha_{rk} > 0$ and

$$\frac{\alpha_{r0}}{\alpha_{rk}} = \min_{i \geqslant 1} \left\{ \frac{\alpha_{i0}}{\alpha_{ik}} \,\middle|\, \alpha_{ik} > 0 \right\} .$$

Using the replacement rules to go from one tableau to its dual, if we have a dual feasible tableau, we pivot in row $k$ $(\geqslant 1)$, where

$$\alpha_{k0} = \min_{j \geqslant 1} \{\alpha_{j0}\}$$

and column $r$ $(\geqslant 1)$ where $\alpha_{kr} < 0$ and

$$\frac{\alpha_{0r}}{-\alpha_{kr}} = \min_{i \geqslant 1} \left\{ \frac{\alpha_{0i}}{-\alpha_{ki}} \,\middle|\, \alpha_{ki} < 0 \right\}.$$

Expressed in words, the pivot row is the one with the most negative RHS, and to find the pivot we look for negative entries in the pivot row for which the ratio of the objective row element in the same column to the absolute value of the entry is minimised. Ties can be broken arbitrarily. To illustrate the method in action we shall solve problem P4.

$$\begin{array}{lll} \text{P4:} & \text{maximise} & -x_1 - x_2 - x_3 \quad (= z) \\ & \text{subject to} & x_1 - 2x_2 - x_3 \leqslant -2 \\ & & -2x_1 + 2x_3 + x_3 \leqslant 3 \\ & & x_1 - x_2 - 2x_3 \leqslant -3 \\ & & x_1, x_2, x_2 \geqslant 0. \end{array}$$

| P4 | $x_1$ | $x_2$ | $x_3$ | T1 |
|---|---|---|---|---|
| $s_1$ | 1 | −2 | −1 | −2 |
| $s_2$ | −2 | 2 | 1 | 3 |
| $s_3$ | 1 | −1 | (−2) | −3 |
| $z$ | 1 | 1 | 1 | 0 |

The initial tableau is P4/T1. P4/T1 is dual but not primal feasible. This means we can choose a pivot according to the dual simplex rules and leads us to pivot in row 3 and column 3. Pivoting, and continuing similarly for another iteration, we get P4/T2 and T3. Hence, the optimal solution is $(x_1, x_2, x_3) = (0, \frac{1}{3}, 1\frac{1}{3})$ and the optimal objective function value is $-1\frac{2}{3}$.

| P4 | $x_1$ | $x_2$ | $s_3$ | T2 |
|---|---|---|---|---|
| $s_1$ | $\frac{1}{2}$ | ($-\frac{3}{2}$) | $-\frac{1}{2}$ | $-\frac{1}{2}$ |
| $s_2$ | $-\frac{3}{2}$ | $\frac{3}{2}$ | $\frac{1}{2}$ | $1\frac{1}{2}$ |
| $x_3$ | $-\frac{1}{2}$ | $\frac{1}{2}$ | $-\frac{1}{2}$ | $1\frac{1}{2}$ |
| $z$ | $1\frac{1}{2}$ | $\frac{1}{2}$ | $\frac{1}{2}$ | $-1\frac{1}{2}$ |

| P4 | $x_1$ | $s_1$ | $s_3$ | T3 |
|---|---|---|---|---|
| $x_2$ | $-\frac{1}{3}$ | $-\frac{2}{3}$ | $\frac{1}{3}$ | $\frac{1}{3}$ |
| $s_2$ | $-1$ | 1 | 0 | 1 |
| $x_3$ | $-\frac{1}{3}$ | $\frac{1}{3}$ | $-\frac{2}{3}$ | $1\frac{1}{3}$ |
| $z$ | $1\frac{2}{3}$ | $\frac{1}{3}$ | $\frac{1}{3}$ | $-1\frac{2}{3}$ |

In trying to select the pivot column it may happen that $\alpha_{ro} < 0$ and $\alpha_{rj} \geqslant 0$ for all $j \geqslant 1$, so there is no pivot allowed by the rules above. In this case, since row $r$ of the tableau represents the equation

$$x_{B_r} + \sum_j \alpha_{rj} x_{N_j} = \alpha_{ro},$$

we see that there can be no solution in which $x_{B_r}, x_{N_j} \geqslant 0$ for all $j \geqslant 1$. This says that the problem is infeasible. We might have deduced this from the fact that, if $\alpha^P_{ri} \geqslant 0$ for all $i \geqslant 1$ and $\alpha^P_{ro} < 0$, then $\alpha^D_{ir} \leqslant 0$ for all $i \geqslant 1$ and $\alpha^D_{ro} < 0$ and this means that D is unbounded. In Section 5.2 we saw that this implies that P is infeasible.

Although the dual simplex method can yield a useful saving in computation time in examples such as P4, it requires an extra backward transformation to obtain the pivot row when the PFRS method is used (replacing the forward transformation to obtain the pivot column and usually entailing more work). So, if its application to problems with an initial tableau which is dual feasible were the only justification for considering the method, it would be likely to remain an interesting curiosity. However, as we shall see in subsequent sections, there are many situations, for example, resource parametrisation and integer programming, where dual feasible tableaux arise naturally and the dual simplex method is invaluable as a means of restoring optimality.

## 5.4 THE DUALITY THEOREM AND SOME CONSEQUENCES

In Section 5.2 we saw that any pair of feasible solutions satisfying the CS conditions are optimal in their respective problems. The converse result, that any optimal pair of solutions of primal and dual problems satisfy the CS conditions, will be true if the optimal objective function values of both problems are the same (since $-w = z$ is equivalent to the CS conditions). This equality is the substance of the duality theorem which states that, *if an LP and its dual problem are both feasible, then they both have the same optimal objective function value.*

Before seeing why this is true, we should first observe that the apparently self-evident nature of the theorem is misleading. The same result can fail to hold, if, for example, we demand of only one variable that it must be integer valued. (See Chapter 11.) Such a failure renders the CS conditions incomplete in that, although it is still true that a pair of feasible solutions of the CS conditions are optimal, the optimal solutions may (indeed, usually will) fail to satisfy them. Thus the duality theorem is a valuable and non-trivial result. It can be proved using the PFRS method (Exercise 8) but we shall obtain it is a consequence of the dual simplex method.

In Section 5.3 we saw that, if the simplex method is applied to an LP with an initial primal feasible tableau and the dual simplex method is applied to its dual problem, a sequence of dual tableaux is generated, terminating in an optimal pair of dual tableaux. This conclusion is true even if the LP is not initially primal feasible, provided it is in the form of problem P of Section 5.1 The two-phase method can be used and no artifical variables will be needed, but pivoting on corresponding elements in the dual problem will again generate a sequence of dual tableaux, since the result that pivoting on corresponding elements preserves duality of tableaux was proved in the last section without reference to pivot selection rules. (Note that the objective function of P must be included as the objective row in Phase I. This means that $z^{\mathrm{I}}$ is only used to determine the pivot, but not explicitly included in the tableau.) If degeneracy is present, some device to prevent cycling, such as the mechanism described in Section 3.6, may be required. However, the precise details are unimportant, we only need the result (from Section 3.6) that there is always some set of pivots which leads to an optimal tableau of P. Note that P is feasible and bounded (because D is feasible). We may conclude that P and D will always possess optimal tableaux forming a dual pair and, in these tableaux, we have $\hat{z} = \alpha_{00}^{P} = -\alpha_{00}^{D} = -\hat{w}$, which is just the duality theorem.

The duality theorem provides a complete optimality test. Given a feasible solution of an LP, we look for a feasible solution of the dual problem so that both solutions satisfy the CS conditions. If we find such a dual solution, the given solution is optimal. The duality theorem says that if the given solution is optimal then any optimal solution of the dual problem satisfies the CS conditions and so failure to find a feasible dual solution satisfying the conditions indicates that the given solution is not optimal.

To illustrate, let us test $(x_1, x_2, x_3, x_4) = (0, 2, 0, 3)$ for optimality in P1 (and suppose we have not already tested $(0, 3\frac{1}{8}, \frac{3}{4}, 0)$). Then we must find a dual solution with $t_2 = t_4 = 0$, (because $x_2, x_4 > 0$) and $y_1 = 0$ (because $s_1 > 0$). So $y_2$ and $y_3$ must be a solution of

$$4y_2 + 2y_3 = 8$$
$$3y_2 + \phantom{2}y_3 = 3,$$

which means $y_2 = -1$, $y_3 = 6$. Since this is not feasible, there is no feasible solution of D1 satisfying the CS conditions, so (0, 2, 0, 3) is not optimal.

So far, we have discussed duality only for problems with inequality constraints. Suppose, however, that an LP includes the constraint

$$\sum_j a_{ij} x_j = b_i,$$

then this may be rewritten as two inequalities:

$$\sum_j a_{ij} x_j \leqslant b_i \tag{5.2a}$$

and

$$\sum_j (-a_{ij}) x_j \leqslant -b_i. \tag{5.2b}$$

If we now imagine that problem P has been augmented by the addition of equality constraints, then we can use this trick to rewrite the problem with inequalities only. If $y_i^+$ and $y_i^-$ are the dual variables of constraints (5.2a) and (5.2b), the dual problem looks, schematically, like

$$\begin{array}{lll} \text{minimise} & \ldots\ldots + b_i\, y_i^+ - b_i\, y_i^- + \ldots & \\ \text{subject to} & \ldots\ldots + a_{ij}\, y_i^+ - a_{ij}\, y_i^- + \ldots \geqslant c_j & \text{for all } j \\ & \ldots\ldots\ y_i^+, y_i^- \ldots \geqslant 0 & \text{for all } i. \end{array}$$

Since $y_i^+$ and $y_i^-$ always occur together in the form 'coefficient $\times$ $(y_i^+ - y_i^-)$' in the objective function and constraints, we can replace $y_i^+ - y_i^-$ by a free variable $y_i$ (cf. Section 1.1). So we can associate a *free* dual variable with every *equality* constraint and, by the symmetry of duality, any free variable is associated with an equality constraint in the dual problem. (We must still have '$\leqslant$' constraints when inequalities occur in a 'maximisation' problem.)

As an example, P5 and D5 below form a dual pair

$$\begin{array}{lrl} \text{P5: maximise} & 3x_1 - 2x_2 - \phantom{2}x_3 & \\ \text{subject to} & -x_1 - 2x_2 + \phantom{2}x_3 & \leqslant 4 \\ & x_1 + 7x_2 - \phantom{2}x_3 & = 8 \\ & 3x_1 - 6x_2 + 2x_3 & \leqslant 3 \\ & x_1 + 3x_2 - \phantom{2}x_3 & = -1 \\ & x_1, x_2 \geqslant 0,\ x_3 \text{ free.} & \end{array}$$

$$\begin{array}{lrl} \text{D5: minimise} & 4y_1 + 8y_2 + 3y_3 - y_4 & \\ \text{subject to} & -y_1 + y_2 + 3y_3 + y_4 & \geqslant 3 \\ & -2y_1 + 7y_2 - 6y_3 + 3y_4 & \geqslant -2 \\ & y_1 - y_2 + 2y_3 - y_4 & = -1 \\ & y_1, y_3 \geqslant 0 & \\ & y_2, y_4 \text{ free.} & \end{array}$$

Because we derived duality for equality constraints by considering equivalent inequalities, the duality theorem can be applied. This means that the CS conditions can still be used, provided they are modified so that no free variable has a complementary variable (since its associated dual constraint is an equality). To illustrate, we will test $(x_1, x_2, x_3) = (0, 2\frac{1}{4}, 7\frac{3}{4})$ for optimality in P5. We see that $s_1$, $s_3 > 0$, confirming feasibility and showing that any dual solution of the CS conditions must satisfy $t_2 = y_1 = y_3 = 0$ ($x_3$ has no complementary variable). From the second and third constraints of D5 we deduce that $y_2 = -1\frac{1}{4}$ and $y_4 = 2\frac{1}{4}$. The negative value of $y_2$ is feasible, since it is a free variable. However, substituting this solution in the first constraint of D5 gives $t_1 = -2$. This is not feasible, so the duality theorem allows us to conclude that $(0, 2\frac{1}{4}, 7\frac{3}{4})$ is not optimal in P5.

In Sections 5.2 and 5.4, we have applied the duality theorem, through the CS conditions, to testing given solutions for optimality. In some cases, where an optimal tableau is degenerate, it may be necessary to solve a subsidiary LP in order to apply the test (see Exercise 3). Nevertheless, the subsidiary LP will usually be easier to solve than the primal problem. Although such applications can prove valuable, the major use of the duality theorem (as opposed to the dual simplex method) is as a theoretical tool. In Chapter 10, we shall see that it implies a fundamental result in game theory. In Chapter 8, we shall apply duality to networks. In Chapter 9, we shall use it to relate single- and multiple-objective LPs and, in Chapter 12, we shall derive from it a characterisation of optimality for quadratic objective functions.

## 5.5 AN INTERPRETATION OF DUAL VARIABLES

We will now return to the problem described in Section 5.1 to see how the thoery we have developed permits us to regard dual variables as shadow prices. In so doing, we will obtain a general characterisation of dual variables.

Let us imagine that an LP with objective function $\sum_j c_j x_j$ has an optimal solution $\hat{x}_1, \hat{x}_2, \ldots$ and that the optimal tableau has $\alpha_{0j} > 0$ for $j \geqslant 1$ (indicating a unique optimal solution and that the optimal dual tableau is not degenerate). In equation (3.6a) we saw that $\alpha_{0j}$ is a linear function of the $c_j$s.

This means that it is possible to make small changes in the $c_j$s whilst retaining $\alpha_{0j} \geqslant 0$ and therefore optimality of $\hat{x}_1, \hat{x}_2, \ldots$ . Hence, if the $c_j$s are not changed too much, the optimal objective function value satisfies $\hat{z} = \sum_j c_j \hat{x}_j$ and, in particular, if just one $c_j$ is changed to $c_j + \Delta c_j$, the resulting change in $\hat{z}$ is given by $\Delta\hat{z} = \hat{x}_j \, \Delta c_j$.

Apply this result to the dual problem, assuming that the LP is not degenerate, and using the duality theorem to equate the optimal values of both problems, we see that $\hat{y}_i = \Delta\hat{z}/\Delta b_i$, where $\hat{y}_i$ is the (unique) optimal value of the $i$th dual variable and $\Delta b_i$ is a small charge in $b_i$ (small enough not to disturb the optimality of the currently optimal basis).

In the production problem of 5.1 this says that the entrepreneur's prices (the dual variables) will be equal to the marginal increase in revenue resulting from a small increase in each raw material. If the market price exceeds the shadow price, the manufacturer will be better off selling the raw material, whereas a smaller market than shadow price implies he should buy more. Thus our interpretation of shadow prices is justified for small changes in $b_i$. This explains again the fact that free goods have a shadow price of zero.

These arguments apply only to small changes. It would be worth while selling *some* of the stock of a free good in P1 at any positive price, but to sell *all* the stock would reduce the revenue to zero. A full investigation of non-marginal changes requires the techniques of parametric programming discussed in Chapter 6. One simple result, however, is easily obtained.

No matter what changes are made to $b_i$ for $i \geqslant 1$, the dual solution $\hat{y}_1, \hat{y}_2, \ldots$ remains feasible in the dual problem. If $-\hat{w}$ is the optimal dual objective function value, then we obtain

$$\hat{z} \leqslant -\hat{w} \leqslant \sum_i b_i \hat{y}_i,$$

since the dual objective function is to be minimised. This inequality is much weaker than our marginal result for $\hat{y}_i$ and does not even use the duality theorem, but it does apply for any $b_i$, $i \geqslant 1$, for which the LP is feasible.

## EXERCISES

1. In each of the subsections of this question you are given an LP and a trial solution. Use the CS conditions to test the solution for optimality.

(a) maximise $x_1 + x_2 + x_3$

subject to

$$\begin{aligned} x_1 - x_2 + x_3 &\leqslant 1 \\ 2x_1 + x_2 - 2x_3 &\leqslant 1 \\ x_1 + x_2 + 2x_3 &\leqslant 4 \\ 4x_1 + x_2 + x_3 &\leqslant 6 \end{aligned}$$

$$x_1, x_2, x_3 \geqslant 0.$$

Test $(x_1, x_2, x_3) = (0, 2\frac{1}{2}, \frac{3}{4})$.

(b) minimise $x_1 - x_2 - x_3$

subject to

$$\begin{aligned} 2x_1 - x_2 + 3x_3 &\geqslant -5 \\ x_1 + 2x_2 &\geqslant 6 \\ 3x_1 - x_2 + 2x_3 &\leqslant 1 \\ x_1 - 5x_2 - x_3 &= -13 \\ x_1, x_2, x_3 &\geqslant 0. \end{aligned}$$

Test $(x_1, x_2, x_3) = (\frac{4}{7}, 2\frac{5}{7}, 0)$.

(c) Same LP as (b). Test $(1, 2\frac{4}{5}, 0)$.

2. Solve the following problem by solving its dual problem graphically and using the CS conditions.

minimise $3x_1 + 5x_2 + 2x_3 + 2x_4$

subject to

$$\begin{aligned} 3x_1 + 2x_2 - 4x_3 \qquad &\leqslant -1 \\ 2x_1 + 3x_2 - x_3 + x_4 &\geqslant 2 \\ x_1, x_2, x_3, x_4 &\geqslant 0. \end{aligned}$$

3. Use the CS conditions to show that, if (0, 0, 2) is an optimal solution of P6, then any optimal solution of the dual problem must satisfy the inequalities labelled P7.

P6: maximise $5x_1 + 3x_2 + 6x_3$

subject to

$$\begin{aligned} 3x_1 + 5x_2 + x_3 &\leqslant 4 \\ -2x_1 + x_2 + 3x_3 &\leqslant 6 \\ 2x_1 - 4x_2 + 3x_3 &\leqslant 8 \\ x_1 - x_2 + x_3 &\leqslant 2 \\ x_1, x_2, x_3 &\geqslant 0. \end{aligned}$$

P7:

$$\begin{aligned} -2y_2 + y_4 &\geqslant 5 \\ y_2 - y_4 &\geqslant 3 \\ 3y_2 + y_4 &= 6 \\ y_2, y_4 &\geqslant 0. \end{aligned}$$

Use Phase I of the simplex method to determine whether P7 has a feasible solution or not and hence test (0, 0, 2) for optimality in P6.

4. Write down the dual problem of

maximise $x_1 + \alpha x_2 + x_3$

subject to

$$\begin{aligned} x_1 + 2x_2 + x_3 &\leqslant 4 \\ 2x_1 - x_2 + x_3 &\leqslant \beta \end{aligned}$$

$$x_1 - x_2 - 3x_3 \leqslant 1$$
$$x_1, x_2, x_3 \geqslant 0.$$

Use the CS conditions to answer the following.

(i) Are there values of $\alpha$ and $\beta$ making $(x_1, x_2, x_3) = (2, \frac{1}{4}, \frac{1}{4})$ optimal?
(ii) Show that, if $(2, 1, 0)$ is optimal, $\alpha = 2$ or $\beta = 3$.

[Hint for (ii). Can the CS conditions be satisfied if $\beta > 3$?]

5. Use the dual simplex method to solve

(i) minimise $26x_1 + 7x_2$

subject to
$$6x_1 + 4x_2 \geqslant 30$$
$$5x_1 + 2x_2 \geqslant 23$$
$$3x_1 + 5x_2 \geqslant 29$$
$$x_1, x_2 \geqslant 0.$$

(ii) minimise $3x_1 + 2x_2$

subject to
$$x_1 - x_2 \geqslant 2$$
$$5x_1 - 3x_2 \leqslant 10$$
$$-x_1 + 6x_2 \geqslant -1$$
$$x_1, x_2 \geqslant 0.$$

6. Deduce the pivot column selection rule of the dual simplex method from the requirement that dual feasibility is to be maintained when pivoting.

(Exercises 7 and 8 assume a knowledge of Section 4.1).

7. Explain how the dual simplex method can be carried out using the format of the PFRS method. You will need an extra backwards transformation to determine the pivot row. Apply the method to Exercise 5(i).

8. Show that $y_i = \pi_i$, where $\pi_i$ is calculated by backwards transformation at the final iteration of the PFRS method applied to problem P of Section 5.1, satisfies the constraints of D. Deduce the duality theorem.

9. Suppose that the feasible region of the constraints

$$\sum_j a_{ij} x_j \leqslant b_i \qquad \text{for } i \geqslant 1; \qquad x_j \geqslant 0 \qquad \text{for } j \geqslant 1$$

is bounded. Let $\mu$ be the minimum value of $\sum_j a_{1j} x_j$ subject to these constraints. Show that, if $b_1$ is replaced by $\mu - 1$ in the constraints, they become infeasible.

Use the duality theorem to deduce that at least one of the problems P and D of Section 5.1 must have an unbounded feasible region unless both are infeasible.

CHAPTER 6

# Sensitivity Analysis

## 6.1 DISCRETE CHANGES

Simply obtaining the optimal solution to a linear programming problem is often far from the end of the story. The data, particularly costs and resource availability, may be estimates and therefore subject to review. Alternatively, these parameters may change over time, and solutions of essentially the same problem at different time periods may be required. Yet again, the optimal solution may be felt to be unsatisfactory and further reflection reveal that the reason for this is that some aspect of the problem has not been modelled adequately. In particular, it may be necessary to add or remove constraints. In this chapter, we will study the effect of changes in the data on the optimal solution. This is called **sensitivity analysis**.

Whole books have been devoted to the topic of sensitivity analysis in linear programming, and in one chapter we can do little more than indicate basic themes. We shall start by examining **discrete changes**, in which the coefficients of the objective function or the RHSs of the constraints assume new values, or constraints are added or removed. Other changes are possible, some of which are covered in the exercises. If we wish to find the new optimal solution we could always solve the problem, *ab initio*, using the simplex method. However, the art of sensitivity analysis is in finding methods of obtaining the optimal solution with less effort, in general, using the optimal tableau of the original problem as starting point. For such an approach to be worth while, we would expect the changes in the data to be 'small' in some sense, so the new optimal tableau is close to the old one. It is impossible to quantify these concepts precisely and the assumption that changes are small is more likely to be valid in large-scale problems (where one coefficient, for example, is less important in relation to the rest of the data) than in the simple examples used for illustration, but the underlying principles are independent of such considerations.

We will illustrate the techniques involved by solving examples but the general principles should emerge sufficiently clearly to be readily applied to any problem. We will start by noting that the optimal tableau for problem P1 is

P1/T1, ignoring for the moment the extra rows (a), (b) and (b′), and the extra resource column (c).

$$\begin{array}{lll} \text{P1:} & \text{maximise} & 5x_1 + 4x_2 + 7x_3 + 3x_4 \qquad (=z) \\ & \text{subject to} & 2x_1 + x_2 + 3x_3 + 2x_4 \leqslant 5 \\ & & 4x_1 + 3x_2 + 5x_3 + 3x_4 \leqslant 9 \\ & & x_1 + x_2 + x_3 + x_4 \leqslant 3 \\ & & x_1, x_2, x_3, x_4 \geqslant 0. \end{array}$$

| | P1 | $x_1$ | $s_2$ | $s_1$ | $x_4$ | T1 | (c) |
|---|---|---|---|---|---|---|---|
| | $x_3$ | $\frac{1}{2}$ | $-\frac{1}{4}$ | $\frac{3}{4}$ | $\frac{3}{4}$ | $1\frac{1}{2}$ | $2\frac{1}{2}$ |
| | $x_2$ | ($\frac{1}{2}$) | $\frac{3}{4}$ | ($-\frac{5}{4}$) | $-\frac{1}{4}$ | $\frac{1}{2}$ | $-1\frac{1}{2}$ |
| | $s_3$ | 0 | $-\frac{1}{2}$ | $\frac{1}{2}$ | $\frac{1}{2}$ | 1 | 2 |
| | $z$ | $\frac{1}{2}$ | $1\frac{1}{4}$ | $\frac{1}{4}$ | $1\frac{1}{4}$ | $12\frac{1}{2}$ | $11\frac{1}{2}$ |
| (a) | $z$ | $-3$ | $1\frac{1}{2}$ | $-\frac{1}{2}$ | $\frac{1}{2}$ | 11 | |
| (b) | $s_4$ | $\frac{1}{2}$ | $-\frac{1}{4}$ | ($-\frac{1}{4}$) | $-\frac{1}{4}$ | $-1\frac{1}{2}$ | |
| (b′) | $s'_4$ | ($-\frac{1}{2}$) | $\frac{1}{4}$ | $\frac{1}{4}$ | $\frac{1}{4}$ | $-\frac{1}{2}$ | |

How does this change if the objective function is replaced by

$$8x_1 + 4x_2 + 6x_3 + 3x_4?$$

The general approach to discrete changes is to first find how the optimal tableau changes when the new data is used. This usually results in a tableau which is primal or dual infeasible and the dual or standard simplex method is used to restore optimality. The special methods of sensitivity analysis are only used in the former step so we shall concentrate on that. Clearly, the new objective function will change the objective row of P1/T1. The new objective row can be calculated in the usual way, using formulae (3.6), and the result is the objective row (a). The tableau is no longer optimal, so we must pivot in the $x_1$-column to restore dual feasibility. The PRS rule selects $r = 2$ and two pivots are required to restore optimality, the optimal solution being $(x_1, x_2, x_3, x_4) = (2\frac{1}{4}, 0, 0, 0)$.

Now let us return to P1 and ask what happens if the additional constraint C: $2x_1 + x_2 + 2x_3 + x_4 \leqslant 2$ (or $2x_1 + x_2 + 2x_3 + x_4 + s_4 = 2, s_4 \geqslant 0$) is included in P1. This will obviously entail an extra row in the tableau and to determine the entries in this row we will consider how the general constraint,

$$\sum_j d_j x_j \leqslant d_0 \quad (\text{or } \sum_j d_j x_j + s = d_0, \quad s \geqslant 0)$$

can be incorporated in a tableau. To achieve this, we substitute the canonical equations into the additional constraint to obtain the equation

$$s + \sum_{j \geqslant 1} \alpha_j x_{N_j} = \alpha_0$$

in which the coefficients are given by

$$\alpha_j = d_{N_j} - \sum_i d_{B_i} \alpha_{ij} \quad \text{for } i \geqslant 1 \quad \text{and} \quad \alpha_0 = d_0 - \sum_i d_{B_i} \alpha_{ij}, \tag{6.1}$$

where $d_{B_i}$ ($d_{N_j}$) is the coefficient of $x_{B_i}$ ($x_{N_j}$) in the additional constraint. This derivation is parallel to that used to obtain (3.6) and the reader is ased to supply the details.

Applying (6.1) to the additional constraint leads to the additional row (b) in P1/T1. With this row added to the tableau, primal feasibility is violated and the dual simplex method is used to restore primal feasibility. Pivoting initially in the $s_1$-column of row (b), two dual simplex iterations are required to reach an optimal tableau. The optimal solution is $(x_1, x_2, x_3, x_4) = (0, 2, 0, 0)$.

A slight modification of the procedure will enable us to add equality constraints. For example, if we had desired to add

$$\text{D:} \quad 2x_1 + x_2 + 2x_3 + x_4 = 2$$

to P1, we could, instead, have added C, but then, as soon as $s_4$ became non-basic (after the first pivot), drop $s_4$ and its column. We have, therefore, treated $s_4$ as if it were an artificial variable. The optimal solution does not change but the optimal tableau has one less column.

If the equality constraint to be added had been

$$\text{D}': \quad 2x_1 + x_2 + 2x_3 + x_4 = 4,$$

then just replacing the RHS of C with 4 would have changed the resource column entry in row (b) of P1/T1 to $\frac{1}{2}$ so no pivoting would take place. In this case, rather than C, we add the constraint

$$\text{C}': \quad 2x_1 + x_2 + 2x_3 + x_4 \geqslant 4$$
$$(\text{or } -2x_1 - x_2 - 2x_3 - x_4 + s_4' = -4, \quad s_4' \geqslant 0)$$

which leads to the additional row (b$'$). The new tableau is not primal feasible and pivoting in the $x_1$-column of row (b$'$) gives an optimal tableau with $(x_1, x_2, x_3, x_4) = (1, 0, 1, 0)$. The general procedure, given an equality constraint, is to add whichever inequality is *not* satisfied by the current solution. This will ensure that the augmented tableau is not primal feasible and a pivot in the added row will make the slack variable non-basic. This variable and its column are then dropped. When the equality is exactly satisfied, either inequality can be used but the optimal solution will not change.

To remove an inequality constraint is straightforward if the slack variable for that constraint is basic; we just drop the corresponding row from the tableau. If the slack variable is non-basic, we can pivot to make it basic and then drop the corresponding row. The choice of pivot row to achieve this can affect the computational effort required to restore optimality and is investigated in Exercise 1.

We will now consider the effect of changing the RHSs of the constraints of P1 and we must first see how to calculate the new resource column in P1/T1. A general formula can be deduced on the assumption that all constraints are of '$\leqslant$' form (as in problem P of Section 5.1), in which case one feasible solution is

$$x_j = 0 \quad \text{for } j \geqslant 1, \quad s_i = b_i \quad \text{for } i \geqslant 1, \quad \text{and } z = 0. \tag{6.2}$$

Let us now set

$$b_{N_j}\,(b_{B_i}) = b_I \text{ if } x_{N_j}\,(x_{B_i}) \text{ is } s_I \text{ for some } I \geqslant 1$$
$$= 0 \quad \text{otherwise}$$

where $x_{N_j}$, $x_{B_i}$ refer to the optimal tableau (or any other tableau under consideration). This says that $b_{N_j}$ is the RHS of the constraint in which $x_{N_j}$ is slack variable or zero if there is no such constraint.

With this new notation, (6.2) can be written

$$x_{N_j} = b_{N_j} \quad \text{for } j \geqslant 1 \quad \text{and} \quad x_{B_i} = b_{B_i} \quad \text{for } i \geqslant 1,$$

and this is a feasible solution. Substitution in the canonical equations gives

$$\alpha_{i0} = \sum_{j\geqslant 1} \alpha_{ij}\, b_{N_j} + b_{B_i} \quad \text{and} \quad \alpha_{00} = \sum_{j\geqslant 1} \alpha_{0j}\, b_{N_j}. \tag{6.3}$$

These results can also be obtained by applying the formula (3.6) to the dual tableau.

If the RHSs of the constraints of P1 are changed for 6, 8 and 3 respectively, the new resource column entries are, for example,

in row 1: $8 \times (-\frac{1}{4}) + 6 \times \frac{3}{4} + 0 = 2\frac{1}{2}$

in row 3: $8 \times (-\frac{1}{2}) + 6 \times \frac{1}{2} + 3 = 2$

in row 0: $8 \times 1\frac{1}{4} \quad + 6 \times \frac{1}{4} = 11\frac{1}{2}$

using (6.3). The complete resource column is shown as (c) in P1/T1. It is not primal feasible, but a dual simplex iteration, pivoting in the $x_2$-row and $s_1$-column, restores optimality and the optimal solution is $(x_1, x_2, x_3, x_4) = (0, 0, 1\frac{3}{5}, 0)$.

When equality constraints are present, the formulae (6.3) cannot be used, since the assumption of inequality only constraints was made in deriving (6.3). Indeed, no comparable formula can be found, since information is lost when the artificial variables introduced in Phase I are dropped as they become non-basic.

This presents no extra problem in the PFRS method since a forward transformation of the new RHSs will generate the new pivot column. For small problems, solved by hand, an alternative technique is available. To illustrate it, we will consider P2 whose optimal tableau is P2/T1.

P2: maximise $x_1 - 2x_2 - 2x_3 + x_4$

subject to

$$2x_1 + x_2 + 2x_3 + 3x_4 = 5$$
$$2x_1 - 2x_2 - x_3 + x_4 = 3$$
$$x_1, x_2, x_3, x_4 \geqslant 0.$$

| P2 | $x_2$ | $x_3$ | T1 | |
|---|---|---|---|---|
| $x_4$ | $\frac{3}{2}$ | $\frac{3}{2}$ | 1 | $-1$ |
| $x_1$ | $-\frac{7}{4}$ | $-\frac{5}{4}$ | 1 | $3\frac{1}{2}$ |
| $z$ | $1\frac{3}{4}$ | $2\frac{1}{4}$ | 2 | $2\frac{1}{2}$ |

What is the new optimal solution if the RHSs of the constraints are changed to 4 and 6 respectively? We can find the new resource column by noting that $x_2 = x_3 = 0$ in P2/T1. Hence $x_1$ and $x_4$ must satisfy

$$\left.\begin{aligned} 2x_1 + 3x_4 &= 4 \\ 2x_1 + x_4 &= 6 \end{aligned}\right\} \qquad \begin{aligned} x_1 &= 3\tfrac{1}{2} \quad x_4 = -1 \\ z &= x_1 + x_4 = 2\tfrac{1}{2}. \end{aligned}$$

This gives the new resource column shown in P2/T1. Once the new resource column has been calculated, restoration of optimality is standard. In particular, P2/T1 shows that the new problem is infeasible. In general, we set all non-basic variables to zero and solve for the remainder. This set of $m$ equations in $m$ variables (assuming $m$ constraints) will always have a unique solution, determining the new resource column.

## 6.2 PARAMETRIC PROGRAMMING

In this section, we will consider how the optimal solution of an LP can be found when the data is a function of a single parameter, $\theta$. Of course, the optimal solution will also depend on $\theta$ but, in the case when $\theta$ appears only in the coefficients of the objective function or only on the RHSs of the constraints, the techniques of the preceding section can be modified to solve the problem. If one of the $a_{ij}$s depends on $\theta$, the problem is much more complicated and we will not discuss it further. When there is more than one parameter, other complications ensue. We will not discuss the general multiparametric problem, but a special, though important and extendable, case will be treated in Chapter 9. To justify the use of parametric methods we refer the reader to the LP for paper recycling discussed in Section 1.1 This involves a parameter $\lambda$ (the proportion of available waste paper that is actually recycled). Since the model is speculative, this parameter is unlikely to be known in advance, so RHS parametric programming could be used to see how the optimal objective function value and solution varied with $\lambda$.

We will start by finding the optimal solution for all values of $\theta$ when the objective function of P1 is changed to

$$5x_1 + (4+\theta)x_2 + 7x_3 + 3x_4.$$

We must first calculate the new objective row and can use the formulae (3.6) for $\alpha_{0j}$ to do this. For example, we find

$$\alpha_{01} = \tfrac{1}{2} \times 7 + \tfrac{1}{2} \times (4+\theta) - 5 = \tfrac{1}{2} + \tfrac{1}{2}\theta$$
$$\alpha_{03} = \tfrac{3}{4} \times 7 - \tfrac{5}{4} \times (4+\theta) - 0 = \tfrac{1}{4} - 1\tfrac{1}{4}\theta.$$

For $\theta = 0$, these values agree with the objective row of P1/T1 and we can augment P1/T1 with an extra row displaying the coefficients of $\theta$. The resulting tableau is P1/T2.

| P1 | $x_1$ | $s_2$ | $s_1$ | $x_4$ | T2 |
|---|---|---|---|---|---|
| $x_3$ | $\frac{1}{2}$ | $-\frac{1}{4}$ | $\frac{3}{4}$ | $\frac{3}{4}$ | $1\frac{1}{2}$ |
| $x_2$ | ($\frac{1}{2}$) | $\frac{3}{4}$ | $-\frac{5}{4}$ | $-\frac{1}{4}$ | $\frac{1}{2}$ |
| $s_3$ | 0 | $-\frac{1}{2}$ | ($\frac{1}{2}$) | $\frac{1}{2}$ | 1 |
| $z$ | $\frac{1}{2}$ | $1\frac{1}{4}$ | $\frac{1}{4}$ | $1\frac{1}{4}$ | $12\frac{1}{2}$ |
| $\theta$ | $\frac{1}{2}$ | $\frac{3}{4}$ | $-1\frac{1}{4}$ | $-\frac{1}{4}$ | $\frac{1}{2}$ |

| P1 | $x_1$ | $s_2$ | $s_3$ | $x_4$ | T3 |
|---|---|---|---|---|---|
| $x_3$ | $\frac{1}{2}$ | ($\frac{1}{2}$) | $-\frac{3}{2}$ | 0 | 0 |
| $x_2$ | $\frac{1}{2}$ | $-\frac{1}{2}$ | $\frac{5}{2}$ | 1 | 3 |
| $s_1$ | 0 | $-1$ | 2 | 1 | 2 |
| $z$ | $\frac{1}{2}$ | $1\frac{1}{2}$ | $-\frac{1}{2}$ | 1 | 12 |
| $\theta$ | $\frac{1}{2}$ | $-\frac{1}{2}$ | $2\frac{1}{2}$ | 1 | 3 |

Tableau P1/T2 is optimal provided the objective row is non-negative:

$$\tfrac{1}{2} + \tfrac{1}{2}\theta \geqslant 0, \quad 1\tfrac{1}{4} + \tfrac{3}{4}\theta \geqslant 0, \quad \tfrac{1}{4} - 1\tfrac{1}{4}\theta \geqslant 0, \quad 1\tfrac{1}{4} - \tfrac{1}{4}\theta \geqslant 0,$$

or

$$\theta \geqslant -1, \quad \theta \geqslant -1\tfrac{2}{3}, \quad \theta \leqslant \tfrac{1}{5}, \quad \theta \leqslant 5$$

and these are all satisfied provided $-1 \leqslant \theta \leqslant \frac{1}{5}$. In general, if $\alpha_{\theta j}$ denotes the entry in the $\theta$ -row and $j$th column, the tableau is optimal provided

$$\alpha_{0j} + \theta\alpha_{\theta j} \geqslant 0 \qquad \text{for all } j \geqslant 1$$

or,

$$\theta \geqslant -\alpha_{0j}/\alpha_{\theta j} \qquad \text{for all } j \geqslant 1, \alpha_{\theta j} > 0$$

and

$$\theta \leqslant -\alpha_{0j}/\alpha_{\theta j} \qquad \text{for all } j \geqslant 1, \alpha_{\theta j} < 0.$$

So, if we put

$$\underline{\theta} = \max_{j \geqslant 1} \left\{ \frac{-\alpha_{0j}}{\alpha_{\theta j}} \middle| \alpha_{\theta j} > 0 \right\}, \tag{6.4a}$$

$$\bar{\theta} = \min_{j \geqslant 1} \left\{ \frac{-\alpha_{0j}}{\alpha_{\theta j}} \middle| \alpha_{\theta j} < 0 \right\}, \tag{6.4b}$$

then the tableau is optimal for $\underline{\theta} \leqslant \theta \leqslant \bar{\theta}$. If $\bar{\theta}$ (or $\underline{\theta}$) is not defined because no $\alpha_{\theta j} > 0$ (or $< 0$) then the upper (or lower) bound is absent.

Returning to our example, to find the optimal solution for all $\theta \geqslant 0$, we imagine starting with $\theta = 0$ and then increasing $\theta$. The BFS of P1/T2 is optimal for $\theta$ up to $\frac{1}{5}$. When $\theta = \frac{1}{5}$, $\frac{1}{4} - 1\frac{1}{4}\theta = 0$, indicating an alternative optimal solution (Section 3.5), which can be found by pivoting in the $s_1$-column. The PRS rule selects the indicated pivot ($s_3$-row) and pivoting results in tableau P1/T3. Note that the $\theta$-row is transformed by the usual rules since the entries are simply coefficients of $\theta$. Applying formulae (6.4) to P1/T3 gives $\underline{\theta} = \frac{1}{5}$ and $\bar{\theta} = 3$. The fact that $\underline{\theta}$ for P1/T3 is the same as $\bar{\theta}$ for P1/T2 reflects the fact that these are alternative optimal tableaux when $\theta = \frac{1}{5}$. For $\bar{\theta} = 3$ in P1/T3, an alternative optimal tableau is found by pivoting in the $s_2$-column. Pivoting leads to tableau P1/T4. Note that although these tableaux are different, the degeneracy of P1/T3 means that the basic solution has not changed.

| P1 | $x_1$ | $x_3$ | $s_3$ | $x_4$ | T4 |
|---|---|---|---|---|---|
| $s_2$ | 1 | 2 | −3 | 0 | 0 |
| $x_2$ | 1 | 1 | 1 | 1 | 3 |
| $s_1$ | 1 | 2 | −1 | 1 | 2 |
| $z$ | −1 | −3 | 4 | 1 | 12 |
| $\theta$ | 1 | 1 | 1 | 1 | 3 |

In P1/T4 $\underline{\theta} = 3$, as expected, and there is no upper bound so that P1/T4 is optimal for all $\theta \geqslant 3$. There is no need for further tableaux and we see that

P1/T2 is optimal for $\theta$ up to $\frac{1}{5}$, P1/T3 between $\frac{1}{5}$ and 3 and P1/T4 for larger values of $\theta$.

| P1 | $x_2$ | $s_2$ | $s_1$ | $x_4$ | T5 |
|---|---|---|---|---|---|
| $x_3$ | $-1$ | $-1$ | 2 | 1 | 1 |
| $x_1$ | 2 | $\frac{3}{2}$ | $-\frac{5}{2}$ | $-\frac{1}{2}$ | 1 |
| $s_3$ | 0 | $-\frac{1}{2}$ | $\frac{1}{2}$ | $\frac{1}{2}$ | 1 |
| $z$ | $-1$ | $\frac{1}{2}$ | $1\frac{1}{2}$ | $1\frac{1}{2}$ | 12 |
| $\theta$ | $-1$ | 0 | 0 | 0 | 0 |

The general procedure for $\theta \geqslant 0$ should now be clear. In each tableau we find $\bar{\theta}$ using (6.4b) ( $\underline{\theta}$ is equal to $\bar{\theta}$ for the previous tableau) and, assuming that there is a unique $j$ (say $j = k$) which achieves the maximum in (6.4b), we pivot in column $k$ and repeat the process. (We will see how to proceed if the maximum is achieved at more than one $j$ in Section 6.3.) If there is no $\bar{\theta}$, there is no upper bound and we are finished. If no pivot is possible because $\alpha_{ik} \leqslant 0$ for all $i$, then the LP is unbounded for all $\theta > \bar{\theta}$ (why?). Very often we are only interested in $\theta$ satisfying $\theta \leqslant U$. In this case we would stop as soon as a tableau is reached in which $\bar{\theta} \geqslant U$.

For $\theta \leqslant 0$, we apply a similar procedure, decreasing $\theta$ from zero and calculating $\underline{\theta}$ in each tableau using (6.4a). In P1/T2 this means $\underline{\theta} = -1$ and we pivot in the $x_1$-column (and $x_2$-row by the PRS rule) to get P1/T5. In P1/T5, there is no lower bound, so the tableau is optimal for all $\theta \leqslant -1$. To summarise the results, the optimal solution is

$$
\begin{aligned}
(x_1, x_2, x_3, x_4) &= (1, 0, 1, 0) && z = 12 && \text{for } \theta \leqslant -1 \\
&= (0, \tfrac{1}{2}, 1\tfrac{1}{2}, 0) && z = 12\tfrac{1}{2} + \tfrac{1}{2}\theta && \text{for } -1 \leqslant \theta \leqslant \tfrac{1}{5} \\
&= (0, 3, 0, 0) && z = 12 + 3\theta && \text{for } \theta \geqslant \tfrac{1}{5}.
\end{aligned}
$$

We now turn to RHS parametrisation and ask for the optimal solution as $\theta$ varies when the RHSs of the constraints of P1 are $5, 9 + \theta$ and 3. The procedure

| P1 | $x_1$ | $s_2$ | $s_1$ | $x_4$ | T6 | $\theta$ |
|---|---|---|---|---|---|---|
| $x_3$ | $\frac{1}{2}$ | $-\frac{1}{4}$ | $\frac{3}{4}$ | $\frac{3}{4}$ | $1\frac{1}{2}$ | $-\frac{1}{4}$ |
| $x_2$ | $\frac{1}{2}$ | $\frac{3}{4}$ | $\boxed{-\frac{5}{4}}$ | $-\frac{1}{4}$ | $\frac{1}{2}$ | $\frac{3}{4}$ |
| $s_3$ | 0 | $\boxed{-\frac{1}{2}}$ | $\frac{1}{2}$ | $\frac{1}{2}$ | 1 | $-\frac{1}{2}$ |
| $z$ | $\frac{1}{2}$ | $1\frac{1}{4}$ | $\frac{1}{4}$ | $1\frac{1}{4}$ | $12\frac{1}{2}$ | $1\frac{1}{4}$ |

is dual to that used when the objective function is parametrised. We first write down the new resource column in P1/T1 using the formula (6.3) for $\alpha_{io}$. This gives P1/T6 in which the $\theta$-column contains the coefficients of $\theta$ in the resource column.

P1/T6 is optimal (primal feasible) provided $1\frac{1}{2} - \frac{1}{4}\theta \geqslant 0$, $\frac{1}{2} + \frac{3}{4}\theta \geqslant 0$, $1 - \frac{1}{2}\theta \geqslant 0$, or $-\frac{2}{3} \leqslant \theta \leqslant 2$. More generally, if $\alpha_{i\theta}$ denotes the $\theta$-column, the tableau is optimal for $\underline{\theta} \leqslant \theta \leqslant \overline{\theta}$, where

$$\underline{\theta} = \max_{j \geqslant 1} \left\{ \frac{-\alpha_{io}}{\alpha_{i\theta}} \,\middle|\, \alpha_{i\theta} > 0 \right\}$$

$$\overline{\theta} = \min_{j \geqslant 1} \left\{ \frac{-\alpha_{io}}{\alpha_{i\theta}} \,\middle|\, \alpha_{i\theta} < 0 \right\}.$$

Increasing $\theta$ from $\theta = 0$, we find that when $\theta = 2$, $1 - \frac{1}{2}\theta = 0$ in the $s_3$-row and pivoting in this row, using the dual-simplex pivot column selection rule to maintain dual feasibility, gives P1/T7 which is optimal for all $\theta \geqslant 2$.

| P1 | $x_1$ | $s_3$ | $s_1$ | $x_4$ | T7 | $\theta$ |
|---|---|---|---|---|---|---|
| $x_3$ | $\frac{1}{2}$ | $-\frac{1}{2}$ | $\frac{1}{2}$ | $\frac{1}{2}$ | 1 | 0 |
| $x_2$ | $\frac{1}{2}$ | $\frac{3}{2}$ | $-\frac{1}{2}$ | $\frac{1}{2}$ | 2 | 0 |
| $s_2$ | 0 | −2 | −1 | −1 | −2 | 1 |
| $z$ | $\frac{1}{2}$ | $2\frac{1}{2}$ | $1\frac{1}{2}$ | $2\frac{1}{2}$ | 15 | 0 |

| P1 | $x_1$ | $s_2$ | $x_2$ | $x_4$ | T8 | $\theta$ |
|---|---|---|---|---|---|---|
| $x_3$ | $\frac{4}{5}$ | $\frac{1}{5}$ | $\frac{3}{5}$ | $\frac{3}{5}$ | $1\frac{4}{5}$ | $\frac{1}{5}$ |
| $s_1$ | $-\frac{2}{5}$ | $-\frac{3}{5}$ | $-\frac{4}{5}$ | $\frac{1}{5}$ | $-\frac{2}{5}$ | $-\frac{3}{5}$ |
| $s_3$ | $\frac{1}{5}$ | $-\frac{1}{5}$ | $\frac{2}{5}$ | $\frac{2}{5}$ | $1\frac{1}{5}$ | $-\frac{1}{5}$ |
| $z$ | $\frac{3}{5}$ | $1\frac{2}{5}$ | $\frac{1}{5}$ | $1\frac{1}{5}$ | $12\frac{3}{5}$ | $1\frac{2}{5}$ |

When $\theta = -\frac{2}{3}$ in P1/T6, $\frac{1}{2} + \frac{3}{4}\theta = 0$ in the $x_2$-row, giving the pivot indicated in that row and resulting in P1/T8, optimal for $-9 \leqslant \theta \leqslant -\frac{2}{3}$. When $\theta = -9$, $1\frac{4}{5} + \frac{1}{5}\theta = 0$ but the dual simplex pivot selection rule does not give a pivot column (because all entries are non-negative). For any $\theta < -9$, $1\frac{4}{5} + \frac{1}{5}\theta < 0$ and this observation implies infeasibility. So the problem is infeasible for $\theta < -9$ and, for $\theta \geqslant -9$, the optimal solution is given by

$$(x_1, x_2, x_3, x_4) = (0, 0, 1\tfrac{4}{5} + \tfrac{1}{5}\theta, 0), \qquad z = 12\tfrac{3}{5} + 1\tfrac{2}{5}\theta \quad \text{for } -9 \leqslant \theta \leqslant \tfrac{2}{3}$$

$$= (0, \tfrac{1}{2} + \tfrac{3}{4}\theta, 1\tfrac{1}{2} - \tfrac{1}{4}\theta, 0), \qquad z = 12\tfrac{1}{2} + 1\tfrac{1}{4}\theta \quad \text{for } -\tfrac{2}{3} \leqslant \theta \leqslant 2$$

$$= (0, 2, 1, 0) \qquad z = 15 \quad \text{for } \theta \geqslant 2.$$

In our examples only one variable contained the parameter $\theta$. This is not essential. If the objective function in P1 had been

$$(5 + 2\theta)x_1 + (4 - 3\theta)x_2 + (7 + \theta)x_3 + (3 + 4\theta)x_4$$

we could have used the formulae (3.6) for $\alpha_{0j}$ to generate the extra row

| $\theta$ | $-3$ | $-2\frac{1}{2}$ | $4\frac{1}{2}$ | $-2\frac{1}{2}$ | $0$ |
|---|---|---|---|---|---|

and then proceed exactly as before. The reader should check that P1/T2 remains optimal for $-\frac{1}{18} \leqslant \theta \leqslant \frac{1}{6}$ and that we pivot in the $x_1$-column when $\theta = \frac{1}{6}$ and the $s_1$-column when $\theta = -\frac{1}{18}$.

For problems with equality constraints subject to RHS parametrisation it may be necessary to use the trick of setting non-basic variables to zero described in Section 6.1. For example, if the RHS of the constraints of P2 had been changed to $5 - 2\theta$ and $3 + \theta$ respectively, we would have to solve

$$2x_1 + 3x_4 = 5 - 2\theta \qquad x_1 = 1 + 1\tfrac{1}{4}\theta \qquad x_4 = 1 - 1\tfrac{1}{2}\theta$$

$$2x_1 + x_4 = 3 + \theta \qquad z = 2 - \tfrac{1}{4}\theta.$$

With this new resource column we proceed exactly as above.

All of the problems considered in this section have had an optimal solution at $\theta = 0$. This need not always be the case. Exercise 6 suggests a method for dealing with the case where the LP is unbounded or infeasible at $\theta = 0$, but not for all values of $\theta$.

## 6.3 FINITENESS OF PARAMETRIC PROGRAMMING

In the examples of parametric programming solved in Section 6.2, we had $\overline{\theta} > \underline{\theta}$ in every tableau. Since $\overline{\theta}$ for one tableau is equal to $\underline{\theta}$ for the next, when increasing $\theta$, this ensures that we cannot repeat a tableau. This strict inequality need not always occur; as we shall see below, tableaux may arise with $\overline{\theta} = \underline{\theta}$. This phenomenon raises the possibility of returning to a previous tableau, leading to the occurrence of cycling (cf. Chapter 3). In this section, we shall describe a method of avoiding cycling, allowing us to conclude that parametric programming problems can be solved in a finite number of steps. Apart from its intrinsic interest, this result will have applications in other chapters.

We will start by looking at the parametric problem P3.

$$
\begin{array}{lllll}
\text{P3:} & \text{maximise} & (4\theta - 8)x_1 + (6\theta - 15)x_2 + (3\theta - 6)x_3 & & \\
 & \text{subject to} & 2x_1 + \quad x_2 - \quad 3x_3 \leqslant 5 & & \\
 & & x_1 + \quad 2x_2 \quad\quad \leqslant 4 & & \\
 & & x_1 + \quad\quad 2x_3 \leqslant 10 & & \\
 & & x_1, x_2, x_3 \geqslant 0. & &
\end{array}
$$

When $\theta = 0$ the initial tableau ($x_1 = x_2 = x_3 = 0$) is optimal. Adding a $\theta$-row gives the tableau P3/T1.

| P3 | $x_1$ | $x_2$ | $x_3$ | T1 |
|---|---|---|---|---|
| $s_1$ | 2 | 1 | −3 | 5 |
| $s_2$ | 1 | 2 | 0 | 4 |
| $s_3$ | 1 | 0 | 2 | 10 |
| $z$ | 8 | 15 | 6 | 0 |
| $\theta$ | −4 | −6 | −3 | 0 |

| P3 | $s_1$ | $x_2$ | $x_3$ | T2 |
|---|---|---|---|---|
| $x_1$ | $\frac{1}{2}$ | $\frac{1}{2}$ | $-\frac{3}{2}$ | $2\frac{1}{2}$ |
| $s_2$ | $-\frac{1}{2}$ | $\frac{3}{2}$ | $\frac{3}{2}$ | $1\frac{1}{2}$ |
| $s_3$ | $-\frac{1}{2}$ | $-\frac{1}{2}$ | $\frac{7}{2}$ | $7\frac{1}{2}$ |
| $z$ | −4 | 11 | 18 | −20 |
| $\theta$ | 2 | −4 | −9 | 10 |

In P3/T1, we have no $\underline{\theta}$, and $\overline{\theta} = 2$. The minimum in (6.4b) is achieved at $j = 1$ and 3 ($x_1$- and $x_3$-columns). If we pivot in the first of these, as indicated in the tableau, we obtain P3/T2 in which $\underline{\theta} = \overline{\theta}$. We will say that a tableau is **optimal over a degenerate interval,** if $\underline{\theta} = \overline{\theta}$. However, we can still apply the rules of Section 6.2. The minimum in (6.4b) is achieved at $j = 3$. This gives the pivot indicated and leads to P3/T3 which is also optimal over a degenerate interval. Following the same rules, we make the indicated pivot to arrive at P3/T4 in which $\overline{\theta} = 5 > \underline{\theta} = 2$. From P3/T4, we proceed as in Section 6.2. One more pivot, as indicated, is required to complete the solution.

| P3 | $s_1$ | $x_2$ | $s_2$ | T3 |
|---|---|---|---|---|
| $x_1$ | 0 | 2 | 1 | 4 |
| $x_3$ | $-\frac{1}{3}$ | 1 | $\frac{2}{3}$ | 1 |
| $s_3$ | $\frac{2}{3}$ | −4 | $-\frac{7}{3}$ | 4 |
| $z$ | 2 | −7 | −12 | −38 |
| $\theta$ | −1 | 5 | 6 | 19 |

| P3 | $s_3$ | $x_2$ | $s_2$ | T4 |
|---|---|---|---|---|
| $x_1$ | 0 | 2 | 1 | 4 |
| $x_3$ | $\frac{1}{2}$ | −1 | $-\frac{1}{2}$ | 3 |
| $s_1$ | $\frac{3}{2}$ | −6 | $-\frac{7}{2}$ | 6 |
| $z$ | −3 | 5 | −5 | −50 |
| $\theta$ | $1\frac{1}{2}$ | −1 | $2\frac{1}{2}$ | 25 |

In P3/T1 we had a choice of pivot column (first or third). We will call these columns and, more generally, any column $j$ for which $\alpha_{0j} + \bar{\theta}\alpha_{\theta j} = 0$ a **critical column**. To be consistent with the procedure of Section 6.2, we must pivot in a critical column. Since $\alpha_{0j} + \bar{\theta}\alpha_{\theta j} = 0$ by definition in a critical column, the objective row evaluated at $\bar{\theta}$ does not change under pivoting. Now $\bar{\theta}$ is equal to $\underline{\theta}$ in the new tableau and, if this tableau is optimal over a degenerate interval, $\bar{\theta}$ is the same for both tableaux. Hence, the set of critical columns will remain the same throughout a sequence of tableaux optimal over degenerate intervals. It is readily verified that the first and third columns are critical in P3/T1, T2, and T3.

It follows from (6.4b) that there will always be a critical column $j$ with $\alpha_{\theta j} < 0$, provided $\bar{\theta}$ is defined. It is possible that there is no pivot in this column because $\alpha_{ij} \leqslant 0$ for all $i \geqslant 1$. But then, since $\alpha_{0j} + \theta\alpha_{\theta j} < 0$ for $\theta > \bar{\theta}$ ($\alpha_{0j} + \bar{\theta}\alpha_{\theta j} = 0$ and $\alpha_{\theta j} < 0$), the problem is unbounded for $\theta > \bar{\theta}$ (see Section 3.4) and the solution is complete. Otherwise, we can pivot in this column. Thus, if $\alpha_{\theta j} > 0$ in all the columns which were critical in the preceding tableau, the current tableau must contain at least one new critical column, which means that the current tableau is optimal over a non-degenerate interval. For example, in P3/T3, we have $\alpha_{\theta 1}, \alpha_{\theta 3} \geqslant 0$ (columns one and three are critical) and we can verify that $\bar{\theta} > \underline{\theta}$ in P3/T4 as asserted (only column two is now critical). When applied to a sequence of tableaux optimal over a degenerate interval, this process can be interpreted as applying the simplex method restricted to the critical columns. This suggests an effective rule, based on the pivot column selection rule of the simplex method, for choosing the pivot column: choose critical column $k$ where $\alpha_{0k} = \min \alpha_{0j}$ and the minimum is over critical columns. If this results in a tie, an arbitrary choice may be made. In P3/T1 we pivoted in the first column because $\alpha_{\theta 1} < \alpha_{\theta 3}$.

The finiteness of the simplex method, established in Chapter 3, shows that in any sequence of tableaux optimal over degenerate intervals, we must eventually reach a tableau with $\alpha_{\theta j} \geqslant 0$ for all critical columns $j$, and consequently, at the next iteration, a tableau optimal over a non-degenerate interval. Since simplex pivots can be chosen to avoid cycling and $\bar{\theta} > \underline{\theta}$ ensures we will not reach the preceding tableau again, we will never repeat a tableau, when increasing $\theta$. Similar results apply to decreasing $\theta$ (choose a column $j$ with $\alpha_{0j} + \underline{\theta}\alpha_{\theta j} = 0$, maximising $\alpha_{\theta j}$ over such columns). Consequently, the parametric programming method is finite.

This result has indirect value as well as direct practical significance. For example we can use it to show that *it is possible to pivot from any primal feasible tableau* (T1) *corresponding to a set of constraints to any other primal feasible tableau* (T2) *for these constraints, maintaining primal feasibility*. We first observe that T1 is uniquely optimal for the objective function $\sum_j -x_{N_j}$ because we would have $\alpha_{0j} = 1 > 0$ for all $j \geqslant 1$. This objective can be rewritten in terms of the structural variables as $\sum_j c_j x_j$, say. Similarly, there is an objective function $\sum_j d_j x_j$ which has T2 as its optimal tableau. Now consider the parametric programming problem with objective function

$$\sum_j (c_j + \theta (d_j - c_j))x_j.$$

When $\theta = 0$, T1 is optimal and, when $\theta = 1$, T2 is optimal. Consequently, we can increase $\theta$ from 0 to 1 and the finiteness of the procedure proves the result. This gives an *a posteriori* justification of the method used to solve Exercise 6 of Chapter 2. A refined version of this argument will be used to validate a form of the simplex method for multiple objectives described in Chapter 9.

We have only dealt with a parametrised objective function so far in the section. However, everything translates into RHS parametrisation. In particular, when increasing $\theta$, we pivot in row $i$, with $\alpha_{i0} + \bar{\theta}\alpha_{i\theta} = 0$, minimising $\alpha_{i\theta}$ over such rows and, when decreasing $\theta$, in row $i$ with $\alpha_{i0} + \underline{\theta}\alpha_{i\theta} = 0$, maximising $\alpha_{i\theta}$ over such rows.

## EXERCISES

1. Find the optimal tableau of P5.

P5: maximise $x_1 + x_2$

subject to

$$\begin{aligned} x_1 - x_2 + x_3 &\leqslant 1 \\ 2x_1 + x_2 - 2x_3 &\leqslant 1 \\ x_1 + x_2 + 2x_3 &\leqslant 4 \\ 4x_1 + x_2 + x_3 &\leqslant 6 \\ x_1, x_2, x_3 &\geqslant 0. \end{aligned}$$

Use the optimal tableau to deduce the new optimal solution when

(i) the objective function is changed to $x_1 + x_3$,
(ii) the right-hand sides are changed from 1, 1, 4, 6 to 0, 1, 3, 5,
(iii) the objective function is changed to $4x_1 + x_2$ *and* the right-hand side of the third constraint is changed to 9,
(iv) the constraint $x_1 + 2x_2 + 3x_3 \leqslant 7$ is added to the problem,
(v) the constraint $3x_1 + 3x_2 + 2x_3 = 8$ is added to the problem,
(vi) the constraint $3x_1 + 3x_2 + 2x_3 = 12$ is added to the problem,
(vii) an extra variable $x_4 \geqslant 0$, is added, so that the constraints become

$$\begin{aligned} x_1 - x_2 + x_3 - x_4 &\leqslant 1 \\ 2x_1 + x_2 - 2x_3 - x_4 &\leqslant 1 \end{aligned}$$

$$x_1 + x_2 + 2x_3 + x_4 \leqslant 4$$
$$4x_1 + x_2 + x_3 + x_4 \leqslant 6,$$

(viii) the first constraint is removed,
(ix) the second constraint is removed,
(x) $x_1$ is removed,
(xi) $x_2$ is removed.

2. For what range of values of $c_1$, the coefficient of $x_1$ in the objective function of P5 (Exercise 1), does the optimal solution found in Exercise 1 remain optimal? Answer this question also for $c_2$ and $c_3$.

3. Find the optimal solution of P5 as a function of $\theta$, when

(i) the objective function is $(1 + \theta)x_1 + x_2$,
(ii) the objective function is $x_1 + (1 + \theta)x_2 + \theta x_3$,
(iii) the RHSs of the constraints are $1 + \theta, 1 + \theta, 4 - \theta, 6 + 2\theta$ respectively.

4. Find the optimal solution of the following problem

(i) when $\phi = 0$ as a function of $\theta$,
(ii) when $\theta = 0$ as a function of $\phi$.

$$\begin{array}{ll} \text{minimise} & -(5+\theta)x_1 + 5x_2 - 4x_3 - (4-3\theta)x_4 + (3+\theta)x_5 \\ \text{subject to} & x_1 + x_2 + 4x_3 + 2x_4 + 5x_5 = 4 - \phi \\ & 7x_1 - x_2 + 4x_3 + 6x_4 + 3x_5 = 12 + 2\phi \\ & x_1, x_2, x_3, x_4, x_5 \geqslant 0. \end{array}$$

5. Find the optimal solution for the following problem as a function of $\theta$

$$\begin{array}{ll} \text{maximise} & (1-\theta)x_1 + (4-\theta)x_2 + x_3 \\ \text{subject to} & x_1 + 2x_2 + x_3 \leqslant 4 + \theta \\ & x_1 - x_2 - x_3 \leqslant 1 - 3\theta \\ & x_1, x_2, x_3 \geqslant 0. \end{array}$$

6.
(i) Use Phase I of the simplex method to find a feasible non-negative solution of the following inequalities

$$2x_1 + x_2 - \theta \leqslant 4$$
$$x_1 + 4x_2 + \theta \leqslant 8$$
$$x_1 + x_2 \geqslant 3.$$

Calling your solution $(\bar{x}_1, \bar{x}_2, \bar{\theta})$, find the optimal tableau of the following LP when $\theta = \bar{\theta}$. Hence, find the optimal solution for all $\theta \geqslant 0$.

$$\begin{array}{ll} \text{maximise} & x_1 + 2x_2 \\ \text{subject to} & 2x_1 + x_2 \leqslant 4 + \theta \\ & x_1 + 4x_2 \leqslant 8 - \theta \\ & x_1 + x_2 \geqslant 3 \\ & x_1, x_2 \geqslant 0. \end{array}$$

Why could you not have just solved the LP with $\theta = 0$?
What could you say if RHS of the second constraint were $4 - \theta$?

(ii) By considering its dual problem, or otherwise, find the optimal solution of the following problem as a function of $\theta \geqslant 0$.

$$\begin{array}{ll} \text{maximise} & (4\theta - 1)x_1 + (2 - 5\theta)x_2 \\ \text{subject to} & x_1 - x_2 \leqslant 1 \\ & 3x_1 - 2x_2 \leqslant 6 \\ & x_1, x_2 \geqslant 0. \end{array}$$

(iii) What could you say if the objective function in (ii) were changed to

$$(2\theta - 1)x_1 + (2 - \theta)x_2?$$

7. Show that the method of Section 6.3 can be adapted to parametrised RHSs by solving the following problem.

$$\begin{array}{ll} \text{minimise} & 3x_1 + 7x_2 \\ \text{subject to} & -x_1 + 2x_2 \leqslant \theta \\ & x_1 - x_2 \leqslant -2\theta \\ & x_1 - 2x_2 \leqslant -3\theta \\ & -2x_1 - x_2 \leqslant 3 \\ & x_1, x_2 \geqslant 0. \end{array}$$

CHAPTER 7

# Bounded Variables

## 7.1 IMPLICIT CONSTRAINTS

One type of constraint, which occurs frequently enough to merit special consideration, is an upper bound on an individual variable. In this chapter, we will see how such constraints can be incorporated implicitly in a tableau. We will start by solving problem P1 by the simplex method.

$$
\begin{array}{lll}
\text{P1:} & \text{maximise} & 7x_1 - x_2 + 6x_3 \\
 & \text{subject to} & x_1 - x_2 - 3x_3 \leqslant 2 \\
 & & 4x_1 - x_2 + 2x_3 \leqslant 5 \\
 & & 0 \leqslant x_1 \leqslant 3, 0 \leqslant x_2 \leqslant 12, 0 \leqslant x_3 \leqslant 2.
\end{array}
$$

Writing $\sigma_1$, $\sigma_2$, $\sigma_3$ for the slack variables in the upper-bound constraints, the solution is given in tableaux P1/T1–T4.

| P1 | $x_1$ | $x_2$ | $x_3$ | T1 |
|---|---|---|---|---|
| $*s_1$ | 1 | −1 | −3 | 2 |
| $*s_2$ | (4) | −1 | 2 | 5 |
| $\sigma_1$ | 1 | 0 | 0 | 3 |
| $\sigma_2$ | 0 | 1 | 0 | 12 |
| $\sigma_3$ | 0 | 0 | 1 | 2 |
| $z$ | −7 | 1 | −6 | 0 |

| P1 | $s_2$ | $x_2$ | $x_3$ | T2 |
|---|---|---|---|---|
| $*s_1$ | $-\frac{1}{4}$ | $-\frac{3}{4}$ | $-\frac{7}{2}$ | $\frac{3}{4}$ |
| $*x_1$ | $\frac{1}{4}$ | $-\frac{1}{4}$ | $\frac{1}{2}$ | $1\frac{1}{4}$ |
| $\sigma_1$ | $-\frac{1}{4}$ | $\frac{1}{4}$ | $-\frac{1}{2}$ | $1\frac{3}{4}$ |
| $\sigma_2$ | 0 | 1 | 0 | 12 |
| $\sigma_2$ | 0 | 0 | (1) | 2 |
| $z$ | $1\frac{3}{4}$ | $-\frac{3}{4}$ | $-2\frac{1}{2}$ | $8\frac{3}{4}$ |

A number of observations of the structure of tableaux P1/T1–T4 can be made. Firstly, if one of $\sigma_J$, $x_J$ is basic and one non-basic then the equation represented by the row containing the basic variable is

$$\sigma_J + x_J = U_J \tag{7.1}$$

where $U_J$ is the upper bound on $x_J$. This follows from the form of the problem and is clearly true in general. In tableau notation, it says that, if $x_{B_r}$ is $x_J$ (or $\sigma_J$) and $x_{N_k}$ is its **partner**: $\sigma_J$ (or $x_J$), then

$$\alpha_{rk} = 1; \quad \alpha_{rj} = 0 \quad \text{for } j \geqslant 1, \quad j \neq k; \quad \alpha_{ro} = U_J. \tag{7.2}$$

| P1 | $s_2$ | $x_2$ | $\sigma_3$ | T3 |
|---|---|---|---|---|
| $*s_1$ | $-\frac{1}{4}$ | $-\frac{3}{4}$ | $\frac{7}{2}$ | $7\frac{3}{4}$ |
| $*x_1$ | $\frac{1}{4}$ | $-\frac{1}{4}$ | $-\frac{1}{2}$ | $\frac{1}{4}$ |
| $\sigma_1$ | $-\frac{1}{4}$ | $\textcircled{\frac{1}{4}}$ | $\frac{1}{2}$ | $2\frac{3}{4}$ |
| $\sigma_2$ | 0 | 1 | 0 | 12 |
| $x_3$ | 0 | 0 | 1 | 2 |
| $z$ | $1\frac{3}{4}$ | $-\frac{3}{4}$ | $2\frac{1}{2}$ | $13\frac{3}{4}$ |

| P1 | $s_2$ | $\sigma_1$ | $\sigma_3$ | T4 |
|---|---|---|---|---|
| $*s_1$ | $-1$ | 3 | 5 | 16 |
| $x_1$ | 0 | 1 | 0 | 3 |
| $*x_2$ | $-1$ | 4 | 2 | 11 |
| $\sigma_2$ | 1 | $-4$ | $-2$ | 1 |
| $x_3$ | 0 | 0 | 1 | 2 |
| $z$ | 1 | 3 | 4 | 22 |

If $\sigma_J$ and $x_J$ are both basic, the sum of the corresponding rows is (7.1). Once again this follows from the fact that (7.1) is just the expression in equality form of the constraint $x_J \leqslant U_J$. In tableau terms, if $x_{B_r}$ is $x_J$ and $x_{B_t}$ is $\sigma_J$, then

$$\alpha_{tj} = -\alpha_{rj} \quad \text{for } j \geqslant 1; \quad \alpha_{to} = U_J - \alpha_{ro}. \tag{7.3}$$

We cannot have both $x_J$ and $\sigma_J$ non-basic, assuming $U_J > 0$, by (7.1).

These results show that, if $x_J$ (or $\sigma_J$) is basic we can deduce the corresponding row from the rest of the tableau. For example, in P1/T1–T4 we could retain only the rows marked with an asterisk in addition to the objective row, without any loss of information. In general, we will not explicitly include any rows for upper-bound constraints, but we will ensure that subsequent tableaux all contain one of $x_J$ or $\sigma_J$, but not both, amongst the basic and non-basic variables. Consequently, we will be able to deduce the implicit rows, corresponding to whichever of $x_J$ or $\sigma_J$ is not contained in the resulting **compact** tableau.

When the pivot column has been selected, the entries in the implicit rows and pivot and resource columns can be deduced from (7.2) and (7.3). The pivot row is determined by the PRS rule and it may be an explicit row, in which case we pivot as usual, or an implicit row, in which case we write down the complete row using (7.2) or (7.3) and pivot using this row. In the latter case, after pivoting, implicit basic variable $x_J$ (or $\sigma_J$) becomes non-basic and the row in which its partner is basic, which may be the pivot or another row, is dropped, to ensure that only one of $\sigma_J$ and $x_J$ is retained in the next tableau. The dropped row will always be of the form (7.2) (zeros and a one in the inner columns).

Compact tableaux can be used in Phase I as well as Phase II. Since $z^{\mathrm{I}}$ only involves slack and artificial variables and, usually, only structural variables are subject to upper bounds, slack and artificial variables do not have associated implicit rows. Consequently the standard formula for the objective row in Phase I applies.

We will illustrate the method by solving P2.

P2: maximise $-2x_1 + 3x_2 + 3x_3 \quad (= z)$

subject to

$$-4x_1 + 3x_2 + 4x_3 \geqslant 9$$
$$-3x_1 + 2x_2 + 3x_3 \leqslant 9$$
$$x_1 \geqslant 0, 0 \leqslant x_2 \leqslant 7, 0 \leqslant x_3 \leqslant 2.$$

The first two tableaux are P2/T1 and T2. In P2/T1 the pivot row is the (implicit) $\sigma_3$-row because

$$2/1 = \min\{9/3\ (s_2\text{-row}),\ 2/1\ (\sigma_3\text{-row})\}$$

and in P2/T2 the pivot row is the $s_2$-row because

$$3/2 = \min\{3/2\ (s_2\text{-row}),\ 7/1\ (\sigma_2\text{-row})\}.$$

Pivoting in P2/T2 gives the first two rows of P2/T3. Phase I is completed.

| P2 | $x_1$ | $x_2$ | $x_3$ | T1 |
|---|---|---|---|---|
| $s_1$ | 4 | −3 | −4 | −9 |
| $s_2$ | −3 | 2 | 3 | 9 |
| $z^{\mathrm{I}}$ | 4 | −3 | −4 | −9 |
| $\sigma_3$ | 0 | 0 | (1) | 2 |

| P2 | $x_1$ | $x_2$ | $\sigma_3$ | T2 |
|---|---|---|---|---|
| $s_1$ | 4 | −3 | 4 | −1 |
| $s_2$ | −3 | (2) | −3 | 3 |
| $z^{\mathrm{I}}$ | 4 | −3 | 4 | −1 |

| P2 | $x_1$ | $s_2$ | $\sigma_3$ | T3 |
|---|---|---|---|---|
| $s_1$ | $-\frac{1}{2}$ | $\frac{3}{3}$ | $-\frac{1}{2}$ | $3\frac{1}{2}$ |
| $x_2$ | $-\frac{3}{2}$ | $\frac{1}{2}$ | $-\frac{3}{2}$ | $1\frac{1}{2}$ |
| $z$ | $-2\frac{1}{2}$ | $1\frac{1}{2}$ | $-1\frac{1}{2}$ | $10\frac{1}{2}$ |
| $\sigma_2$ | ($\frac{3}{2}$) | $-\frac{1}{2}$ | $\frac{3}{2}$ | $5\frac{1}{2}$ |

| P2 | $\sigma_2$ | $s_2$ | $\sigma_3$ | T4 |
|---|---|---|---|---|
| $s_1$ | $\frac{1}{3}$ | $\frac{4}{3}$ | 0 | $5\frac{1}{3}$ |
| $x_1$ | $\frac{2}{3}$ | $-\frac{1}{3}$ | 1 | $3\frac{2}{3}$ |
| $z$ | $1\frac{2}{3}$ | $\frac{2}{3}$ | 1 | $19\frac{2}{3}$ |

When calculating the objective row in P2/T3, we note that

$$x = -2x_1 + 3x_2 - 3\sigma_2 + 6 \qquad (\text{since } x_2 = 2 - \sigma_2)$$

and then use the usual formulae. We can also obtain $\alpha_{00}$ by evaluating $z$ at the BFS of P2/T3: $(x_1, x_2, x_3) = (0, 1\frac{1}{2}, 2)$. In P2/T3 the pivot row is the (implicit) $\sigma_2$-row because $\frac{3}{2}$ in that row is the only positive element in the $x_1$-column.

Since $\sigma_2$ becomes non-basic after pivoting, we drop the $x_2$-row and substitute the $x_1$-row (pivot row in P2/T3) to get P2/T4. This is optimal and the optimal solution has $x_1 = 3\frac{2}{3}, \sigma_2 = \sigma_3 = 0$, which means $(x_1, x_2, x_3) = (3\frac{2}{3}, 7, 2)$.

To show that pivoting in a compact tableau is just an application of the methods of Chapter 2, we wrote out the pivot row explicitly. This imposes no real burden in small problems solved by hand. However, it is unnecessary. If the basic variable in an implicit row is $\sigma_J$ (or $x_J$) and its partner is non-basic, so we pivot on 1 (as in P2/T1), the pivot column $k$ is simply multiplied by $-1$ and the resource column changes to $\alpha'_{i0} = \alpha_{i0} - \alpha_{ik} U_J$. If the partner of $\sigma_J$ (or $x_J$) is basic (as in P2/T3), it is readily verified (see Exercise 3) that the next tableau can also be obtained by pivoting in the row containing the partner of $\sigma_J$ (or $x_J$) and the same pivot column, and multiplying the new pivot column by $-1$. The new non-basic variable must be labelled correctly: $\sigma_J$ (or $x_J$). This is readily verified in P2/T3 using the $x_2$-row and $x_1$-column for pivoting. With these modified rules it is easy to adapt the procedure of this section for use in the PFRS format (Exercise 3).

## 7.2 SENSITIVITY ANALYSIS

A particularly important use of compact tableaux occurs in the solution of integer programming problems (see Chapter 11). In the course of solving such problems we will frequently wish to alter upper bounds and to impose and alter lower bounds on variables. It is therefore important to be able to perform discrete sensitivity analysis and to handle lower bounds. In this section we will consider techniques for solving these problems and, in passing, illustrate how dual simplex iterations may be carried out using compact tableaux.

To illustrate the addition or alteration of upper bounds we will start by adding the constraint $x_1 \leqslant 2$ to problem P2 of Section 7.1. Since $x_1$ is basic in the optimal tableau P2/T4, we can imagine that $x_1$ acquires an implicit row and, since $2 < 3\frac{2}{3}$, this means that $\sigma_1 < 0$. (Adding $x_1 \leqslant U_1$, where $U_1 \geqslant 3\frac{2}{3}$, does not change the solution.) The $\sigma_1$-row is written explicitly in P2/T5; its resource column entry is $2 - 3\frac{2}{3} = -1\frac{2}{3}$.

| P2 | $\sigma_2$ | $s_2$ | $\sigma_3$ | T5 | P2 | $\sigma_2$ | $s_2$ | $\sigma_1$ | T6 | (a) |
|---|---|---|---|---|---|---|---|---|---|---|
| $s_1$ | $\frac{1}{3}$ | $\frac{4}{3}$ | 0 | $5\frac{1}{3}$ | $s_1$ | $\frac{1}{3}$ | $\frac{4}{3}$ | 0 | $5\frac{1}{3}$ | (4) |
| $x_1$ | $\frac{2}{3}$ | $-\frac{1}{3}$ | 1 | $3\frac{2}{3}$ | $x_3$ | $-\frac{2}{3}$ | $\frac{1}{3}$ | 1 | $\frac{1}{3}$ | (3) |
| $z$ | $1\frac{2}{3}$ | $\frac{2}{3}$ | 1 | $19\frac{2}{3}$ | $z$ | 1 | 1 | 1 | 18 | (14) |
| $\sigma_1$ | $-\frac{2}{3}$ | $\frac{1}{3}$ | $(-1)$ | $-1\frac{2}{3}$ | $\sigma_3$ | $\frac{2}{3}$ | $-\frac{1}{3}$ | $(-1)$ | $1\frac{2}{3}$ | $(-1)$ |

Since P2/T5 is primal infeasible, a dual simplex iteration is necessary to restore primal feasibility. By the usual pivot column selection rule, we pivot in the $\sigma_3$-column to obtain P2/T6 (ignore numbers in brackets, *pro tem.*). Note that in P2/T6 we would expect to have $\sigma_3$ basic in the second row. However, for reasons which will appear later, we have replaced the $\sigma_3$-row with the implicit $x_3$-row (making the $\sigma_3$-row implicit). In general, whenever both $x_J$ and $\sigma_J$ are basic, we will retain $x_J$ in the tableau. To verify that P2/T6 really is primal feasible, we must check the resource column entries of implicit rows and, if any is negative, perform further dual simplex iterations. In fact P2/T6 *is* primal feasible, so the new optimal solution is $(x_1, x_2, x_3) = (2, 7, \frac{1}{3})$.

We will now consider the further effect of reducing the upper bound on $x_2$ by 4. (The upper-bound constraints are now $x_1 \leqslant 2$, $x_2 \leqslant 3$ and $x_3 \leqslant 2$.) We must first calculate the effect of this change on the resource column. This can be done by apply the formulae of Chapter 6 or, directly, by noting that replacing $x_2 + \sigma_2 = 7$ with $x_2 + \sigma_2 = 3$ can be effected by replacing $\sigma_2$ with $\sigma_2 + 4$, so that the equation corresponding to the first row: $s_1 + \frac{1}{3}\sigma_2 + \frac{4}{3}s_2 = 5\frac{1}{3}$ becomes $s_1 + \frac{1}{3}(\sigma_2 + 4) + \frac{4}{3}s_2 = 5\frac{1}{3}$ which is equivalent to replacing the RHS with 4. In general, this shows that, if $x_{N_k}$ is $\sigma_J$ and $U_J$ is reduced by $\Delta$, then $\alpha_{i0}$ must be replaced with $\alpha_{i0} - \Delta\alpha_{ik}$. This gives the numbers in brackets in P2/T6. Such a change renders the tableau primal infeasible, since $\sigma_3 < 0$ (because $x_3 > 2$). A dual simplex iteration returns to P2/T5 except for the resource column, which has $s_1 = 4, x_1 = 1$ and $z = 13$. It is primal feasible and therefore optimal, with solution $(x_1, x_2, x_3) = (1, 3, 2)$.

To impose a lower bound $x_J \geqslant L_J$, we need only substitute $x_J = x_J' + L_J$, where $x_J' \geqslant 0$, throughout the problem and solve in the usual way. When upper and lower bounds are imposed on the same variable it is important to note that $x_J \leqslant U_J$ means $x_J' \leqslant U_J - L_J$. (We must have $L_J \leqslant U_J$.)

To illustrate, we will add the bounds $x_2 \leqslant 2$ and $x_3 \geqslant 1$ to P2. This means substituting $x_3' + 1$ for $x_3$ in P2/T6 and gives P2/T7. The terms in brackets indicate changed values and variables.

| | P2 | $\sigma_2$ | $s_2$ | $\sigma_1$ | T7 |
|---|---|---|---|---|---|
| | $s_1$ | $\frac{1}{3}$ | $\frac{4}{3}$ | 0 | $5\frac{1}{3}$ |
| $(x_3')$ | $x_3$ | $\left(-\frac{2}{3}\right)$ (circled) | $\frac{1}{3}$ | 1 | $\frac{1}{3}(-\frac{2}{3})$ |
| | $z$ | 1 | 1 | 1 | 18 |

| P2 | $x_3'$ | $s_2$ | $\sigma_1$ | T8 |
|---|---|---|---|---|
| $s_1$ | $\frac{1}{2}$ | $\frac{3}{2}$ | $\frac{1}{2}$ | $5(4\frac{1}{2})$ |
| $x_2$ | $\frac{3}{2}$ | $\frac{1}{2}$ | $\frac{3}{2}$ | $6(7\frac{1}{2})$ |
| $z$ | $1\frac{1}{2}$ | $1\frac{1}{2}$ | $2\frac{1}{2}$ | $17(15\frac{1}{2})$ |
| $\sigma_2$ | $\frac{3}{2}$ | $-\frac{1}{2}$ | $\left(-\frac{3}{2}\right)$ (circled) | $1(-\frac{1}{2})$ |

Pivoting as shown in P2/T7 gives the optimal tableau P2/T8, in which $x_2 = 6$. $x_3' = 0, \sigma_1 = 0$ or $(x_1, x_2, x_3) = (2, 6, 1)$.

If we now increase the lower bound on $x_3$ from 1 to 2 (meaning $x_3 = 2$, in view of the upper bound on $x_3$), this means we replace $x_3 = x_3' + 1$ with $x_3 = x_3' + 2$, in other words $x_3'$ with $x_3' + 1$. This results in the resource column shown in brackets in P2/T8. The tableau is no longer primal feasible and one further pivot, as shown, is required to achieve an optimal tableau in which $x_1 = 1\frac{2}{3}, \sigma_2 = 0, x_3' = 0$ or $(x_1, x_2, x_3) = (1\frac{2}{3}, 7, 2)$.

## EXERCISES

1. Solve the following problems, using implicit constraints.

(i)
$$\begin{aligned} \text{maximise} \quad & 5x_1 + 2x_2 - 2x_3 \\ \text{subject to} \quad & 4x_1 + x_2 - 2x_3 \leqslant 14 \\ & 2x_1 + x_2 + 2x_3 \leqslant 22 \\ & x_1 + x_2 \leqslant 12 \\ & 0 \leqslant x_1 \leqslant 3,\ 0 \leqslant x_2 \leqslant 8,\ x_3 \geqslant 0. \end{aligned}$$

(ii)
$$\begin{aligned} \text{maximise} \quad & 2x_1 + x_2 + 9x_3 + 3x_4 + 4x_5 \\ \text{subject to} \quad & x_1 + 3x_2 + x_3 - 2x_4 + 2x_5 = 16 \\ & -x_1 + 2x_2 - x_3 + x_4 + 2x_5 = 12 \end{aligned}$$
$$1 \leqslant x_1 \leqslant 3,\ 2 \leqslant x_2 \leqslant 4,\ -2 \leqslant x_3 \leqslant 4,\ -1 \leqslant x_4 \leqslant 4,\ 1 \leqslant x_5 \leqslant 2.$$

2. Use the methods of sensitivity analysis to find the optimal solution when the problem of Exercise 1(i) is modified by

(a) setting $U_3 = 2$ to give $0 \leqslant x_1 \leqslant 3, 0 \leqslant x_2 \leqslant 8, 0 \leqslant x_3 \leqslant 2$,
(b) decreasing $U_2$ to 1 in (a) to give $0 \leqslant x_1 \leqslant 3, 0 \leqslant x_2 \leqslant 1, 0 \leqslant x_3 \leqslant 2$,
(c) setting $L_3 = 6$ to give $0 \leqslant x_1 \leqslant 3, 0 \leqslant x_2 \leqslant 8, x_3 \geqslant 6$,
(d) increasing $L_3$ to 12 in (c) to give $0 \leqslant x_1 \leqslant 3, 0 \leqslant x_2 \leqslant 8, x_3 \geqslant 12$.

3. (Assumes familiarity with Section 4.1.)
Show that pivoting in column $k$ in an implicit row in which $x_J$ (or $\sigma_J$) is basic and replacing the $\sigma_J(x_J)$-row with the new pivot row gives the same tableau as pivoting in column $k$ and the (explicit) $\sigma_J$ (or $x_J$)-row, multiplying the new column $k$ by $-1$ and changing the new $x_{N_k}$ to $x_J$ (or $\sigma_J$). Use this result to show how implicit constraints can be incorporated into the PFRS method and use the method to solve Exercise 1(i).

CHAPTER 8

# Transhipment and Transportation Problems

## 8.1 TRANSHIPMENT PROBLEMS – THE NATURE OF THE BASIC FEASIBLE SOLUTIONS

A large retail organisation owns $m$ sites, some of which may be warehouses and some may be shops. For $i = 1, \ldots, m$, we will write $b_i$ for the excess of requirements over stocks held at site $i$ of some commodity. If the site is a warehouse, we may expect $b_i < 0$, whereas if the site is a shop, we may have $b_i > 0$ and there may be intermediate sites at which no stocks are held or required, so that $b_i = 0$. Our aim is to transport goods from site to site so that requirements are fulfilled from the stocks available at other sites. If we write $x_{ij}$ for the quantity of goods transferred from site $i$ to site $j$ $(i \neq j)$, we are faced with the constraints, for each $i = 1, \ldots, m$,

$$\sum_{j \neq i} x_{ji} - \sum_{k \neq i} x_{ik} = b_i, \tag{8.1}$$

since the first sum represents the flow of goods into $i$ and the second is the flow out of $i$, so that the difference is the net flow into $i$, which must equal the net requirements.

If we sum these equations, we obtain

$$\sum_i b_i = \sum_i \sum_{j \neq i} x_{ji} - \sum_i \sum_{k \neq i} x_{ik} = 0$$

so the $b_i$s must sum to zero if the constraints are to be feasible and we will assume this to be true hereafter. With this assumption, we can drop one of the $m$ constraints without affecting the set of solutions, leaving only $m - 1$ constraints.

Now suppose that the cost of transporting one unit from $i$ to $j$ $(i \neq j)$ is $c_{ij}$ $(\geqslant 0)$ and our objective is to achieve the transfer of goods at minimal cost. Then we must solve the **transhipment problem** (TRP).

$$\text{TRP:} \quad \text{minimise} \quad \sum_i \sum_{j \neq i} c_{ij} x_{ij} \qquad (= -z)$$

$$\text{subject to} \quad \sum_{j\neq i} x_{ji} - \sum_{k\neq i} x_{ik} = b_i \qquad \text{for } i \geqslant 2$$

$$x_{ij} \geqslant 0 \qquad \text{for } i \neq j.$$

Although the TRP is much more specialised than the LPs we have studied so far, a surprisingly wide range of problems can be modelled as TRPs including many examples which have nothing to do with shipping goods between sites. The assignment problem of Section 8.5 and the inventory problem of Exercise 7 provide examples. We shall therefore adopt the neutral term **node** instead of 'site' from now on.

In Section 8.2 we will see that the simplex method can be performed very efficiently for TRPs if we take the structure of the problem into account and we will start by examining the nature of BFSs.

Any BFS must have $m - 1$ basic variables and we will use these variables to establish a relationship between the nodes. We will assign any node $j$ for which $x_{1j}$ or $x_{j1}$ is basic to the first **generation** and say that the **parent** of $j$ is 1, writing $P(j) = 1$. For example, if $m = 9$ and the basic variables are $x_{19}, x_{23}, x_{26}, x_{29}, x_{41}, x_{59}, x_{87}$ and $x_{97}$, then nodes 4 and 9 are assigned to the first generation and $P(4) = P(9) = 1$. Now for each node $i$ assigned to the first generation, we assign any node $j$ ($\neq 1$), not already assigned, for which $x_{ij}$ or $x_{ji}$ is basic, to the second generation and set $P(j) = i$ ($i$ is the parent of $j$). In our example, $x_{29}, x_{59}$ and $x_{97}$ are basic and so nodes 2, 5 and 7 are assigned to the second generation and $P(2) = P(5) = P(7) = 9$. No node has 4 as parent. In general, for a node $i$ of the $r$th generation, we assign any node $j$ ($\neq 1$) not assigned to generations $1, \ldots, r$, for which $x_{ij}$ or $x_{ji}$ is basic, to generation $r + 1$ and set $P(j) = i$. This means that all the remaining nodes in our example are assigned to the third generation and $P(3) = P(6) = 2, P(8) = 7$. It is not clear *a priori* whether we will be able to assign a generation to every node except 1, but we will see subsequently that this is the case for any BFS.

A graphical representation of the relationships generated in our example is given in Fig. 8.1(a). Node $i$ is indicated by a number $i$ in a circle and node $i$ is joined to node $j$ with an (unbroken) line or **edge** and an arrow directed from $i$ to $j$, if $x_{ij}$ is basic. (The broken line should be ignored, *pro tem*.) Nodes belonging to the same generation share the same horizontal level. It may prove helpful to view the figure (ignoring arrows) as a family tree in which the nodes correspond to male members of a family and node 1 is a patriarch. Our earlier terminology ('generation', 'parent') was inspired by this interpretation.

For every node $i \neq 1$ in Fig. 8.1(a) there is a unique **path**, or sequence of distinct nodes linked by edges, from $i$ to 1 (corresponding to 'a line of descent' from 1 to $i$) and this is the defining characteristic of a **tree**. The path can be determined by taking $i$, then $P(i)$, then $P(P(i))$ and so on until 1 is reached. With $i = 6$, this gives $P(6) = 2, P(2) = 9, P(9) = 1$ so the path is $6 \rightarrow 2 \rightarrow 9 \rightarrow 1$. The complete tree (but not the directions of the arrows) can be generated from the

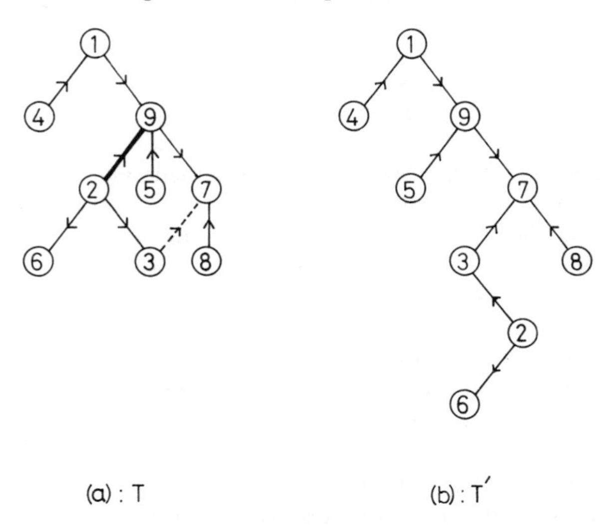


Fig. 8.1 Trees and pivoting in TRPs.

function $P$ as can $P$ from the tree. The trees are a valuable visual aid when solving small problems by hand but 'pointers' such as $P(i)$ are essential for handling problems of a realistic size on a computer.

We will now consider how pivoting may be carried out in the framework we have outlined. Let us suppose that we have a BFS and an associated tree $T$ and that we wish to insert $x_{IJ}$, currently non-basic, into the basis. Since $x_{IJ}$ is non-basic there is currently no edge from $I$ to $J$ but we can always reach $I$ from $J$ in $T$ because we can go from $J$ to 1 and then from 1 to $I$. For example, if we take $I = 3$ and $J = 7$ in Fig. 8.1(a), the unique paths from 7 to 1 and 3 to 1 are $7 \to 9 \to 1$ and $3 \to 2 \to 9 \to 1$ and we can go from 7 to 9 to 1 to 9 to 2 to 3. This is not a path because node 9 is repeated but it can be shortened to the path $7 \to 9 \to 2 \to 3$ by leaving out the redundant section 9 to 1 to 9. In general, we find paths from $J$ to 1 and $I$ to 1 and then look for the first node $k$ at which the paths meet ($k = 9$ in the example). In family-tree terminology, $k$ is the most recent common ancestor of $I$ and $J$. The path from $J$ to $I$ is then constructed by combining the paths from $J$ to $k$ and $k$ to $I$. There are no other paths from $J$ to $I$ for otherwise there would be at least two paths from $J$ to 1 (what are they?).

If we add an edge from $I$ to $J$, we will create a **cycle** (a sequence of distinct edges, each having a node in common with its predecessor, and each node appearing in exactly two edges) by adding the edge $I$ to $J$ to the edges of the path from $J$ to $I$. Furthermore, the cycle is unique because of the uniqueness of the path from $J$ to $I$. If we traverse the cycle in the direction $J$ to $k$ to $I$ to $J$, then some edges of the cycle will have their arrows directed in this direction and we will say that the corresponding variables are **forward** variables whereas the remaining edges will be directed against the direction of the cycle and we will describe the corresponding variables as **reverse** variables. In our example, we have added the edge from 3 to 7 as a dashed line in Fig. 8.1(a) and the resulting cycle consists of the edges from 7 to 9, 9 to 2, 2 to 3 and 3 to 7. This makes $x_{29}$ and $x_{97}$ reverse variables and $x_{23}$ and $x_{37}$ forward variables. In terms of the $P$ function, we note that $x_{IJ}$ is always a forward variable and that $x_{pq}$ is a forward variable if $p$ and $q$ are linked by an edge, either on the path from $J$ to $k$ with $P(p) = q$, or on the path from $I$ to $k$ with $P(q) = p$. The variables corresponding to the remaining edges on these paths will be reverse variables.

Now suppose we increase all forward variables by $\theta$, decrease all reverse variables by $\theta$ and leave all other variables unchanged. Then this new solution satisfies the constraints (8.1) for any $\theta$. To see this, we have only to verify that (8.1) is still satisfied when $i$ is a node of the cycle (why?) so suppose that $s$ to $i$ and $i$ to $t$ are consecutive edges of the cycle. There are four cases to consider.

(i) $x_{si}$ and $x_{it}$ are both forward variables;
(ii) $x_{si}$ is a forward variable and $x_{it}$ is a reverse variable;
(iii) $x_{si}$ is a reverse variable and $x_{it}$ is a forward variable;
(iv) $x_{si}$ and $x_{it}$ are reverse variables.

These cases are illustrated in Fig. 8.2, together with the changes in the values of $x_{si}$ and $x_{it}$. It is clear that the net change in $\sum_{j \neq i} x_{ji} - \sum_{k \neq i} x_{ik}$ is zero. So (8.1) is still satisfied.

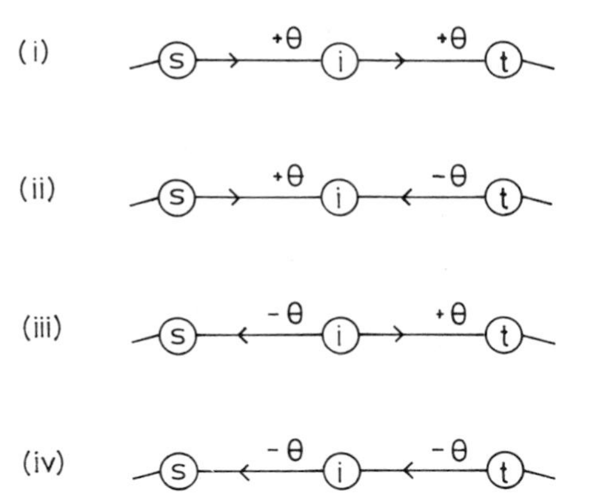


Fig. 8.2 The effect of increasing forward variables and decreasing reverse variables by $\theta$.

If we write $\phi_{ij}$ for the current value of $x_{ij}$ (it will be zero unless $x_{ij}$ is basic), we have seen that, when $x_{IJ} = \theta$, we can still satisfy (8.1) with $x_{ij} = \phi_{ij} - \delta_{ij}\,\theta$

where $\delta_{ij} = 1$ for reverse, $-1$ for forward and zero for all other variables. Consequently, $\delta_{ij}$ is the coefficient of $x_{IJ}$ in the equation of the canonical form in which $x_{ij}$ is basic. To apply the PRS rule, we must choose the smallest $\phi_{ij}/\delta_{ij}$ for $\delta_{ij} > 0$. But $\delta_{ij} > 0$ means that $\delta_{ij} = 1$ and that $x_{ij}$ is a reverse variable, so $x_{pq}$ leaves the basis, where $x_{pq}$ is a reverse variable and

$$\phi_{pq} = \min \{ \phi_{ij} \mid x_{ij} \text{ is a reverse variable} \}.$$

Ties for minimum can be broken arbitrarily. The new BFS is $x_{ij} = \phi_{ij} - \delta_{ij}\phi_{pq}$. If, in our example, we have $\phi_{23} = 5$, $\phi_{29} = 3$, $\phi_{97} = 6$ (the other values are irrelevant at the moment), then

$$\phi_{29} = \min \{ \phi_{29}, \phi_{97} \} = 3$$

since $x_{29}$ and $x_{97}$ are the reverse variables of the cycle created when edge 3 to 7 is added in Fig. 8.1(a). Hence, $x_{29}$ leaves the basis.

Dropping $x_{pq}$ from the basis means deleting the edge from $p$ to $q$ from the tree, $T$, and adding $I$ to $J$. The result is still a tree because, if the path from $i$ to 1 in $T$ is broken by dropping the edge from $p$ to $q$, another path can be constructed by going instead from $p$ to $q$ round the cycle (using the edge $I$ to $J$). Since there is only one cycle created when $I$ to $J$ is added there will be no cycles left after the edge $p$ to $q$ is dropped and so there cannot be more than one path from $i$ to 1 for any $i$. The new tree $T'$ is illustrated in Fig. 8.1(b) for the example studied above.

To handle the change of tree computationally, we must specify how the parent function $P'$ of $T'$ may be determined from $P$. Experimentation with a few examples will show that the following rules have the desired effect. If $p$ and $q$ are on the path from $I$ to $k$ (in which case $q = P(p)$, since $x_{pq}$ is a reverse variable), let $i_1 \to i_2 \to \ldots \to i_s$ be the part of this path from $I$ to $p$, so that $i_1 = I$, $i_s = p$ and $i_{t+1} = P(i_t)$. Then set $P'(i_t) = i_{t-1}$ for $t = 2, \ldots, s$, $P'(I) = J$ and $P'(i) = P(i)$ for all remaining $i \neq 1$. Intuitively, we 'reverse' the path from $I$ to $p$. Similarly, if $p$ and $q$ are on the path from $J$ to $k$, let $i_1 \to i_2 \to \ldots \to i_s$ be the part of this path from $I$ to $q$ so that $i_1 = J$, $i_s = q$ and $i_{t+1} = P(i_t)$. Reverse the path from $J$ to $q$ by setting $P'(i_t) = i_{t-1}$ for $t = 2, \ldots, s$ and put $P'(J) = I$ and $P'(i) = P(i)$ for all remaining $i \neq 1$. In our example, $I = 3$, $J = 7$, $p = 2$, $q = 9$ so that $p, q$ are on the path from $I$ to $k$ $(= 9)$ $i_1 = 3$, $i_2 = 2$ $(s = 2)$. Hence $P'(2) = 3$, $P'(3) = 7$ and for all other $i$ apart from 1, $P'(i) = P(i)$. The reader should check that $P'$ defines the tree $T'$ of Fig. 8.1(b). If, in Fig. 8.1(b), we were to take $I = 4$, $J = 6$, $p = 9$, $q = 7$, putting $p$ and $q$ on the path from $J$ to $k$ $(= 1)$ and giving $i_1 = 6$, $i_2 = 2$, $i_3 = 3$, $i_4 = 7$, the new tree would be defined by $P''(7) = 3$, $P''(3) = 2$, $P''(2) = 6$, $P''(6) = 4$ and $P''(i) = P'(i)$ otherwise $(i \neq 1)$. In applying this method, it is possible that $I = p$ (or $J = q$) in which case we only have to set $P'(I) = J$ or $(P'(J) = I)$.

We have seen that, when pivoting from a BFS corresponding to a tree, the new BFS also corresponds to a tree. In Section 6.3, we showed that it is possible to pivot from any BFS to any other BFS through a sequence of BFSs. Consequently, if we can exhibit one BFS corresponding to a tree, *all* BFSs will correspond to trees. But, $x_{1i} = b_i$, if $b_i \geqslant 0$, and $x_{i1} = -b_i$, if $b_i < 0$, for $i \geqslant 2$ is a BFS. Indeed, if the $i$th equation of TRP is multiplied by $-1$ when $b_i < 0$, the constraints are in canonical form. The tree corresponding to this basis has $P(i) = 1$ for all $i \geqslant 2$.

## 8.2 THE SIMPLEX METHOD FOR THE TRANSHIPMENT PROBLEM

In order to apply the simplex method to the TRP, we must introduce the objective function into the framework established in Section 8.1. In particular, we need to calculate the objective-row coefficient of any non-basic variable. To see how this can be done, we will start by observing that $x_{ij}$ occurs in the $i$th equation of (8.1) with coefficient $-1$ and in the $j$th equation with coefficient $+1$. This means that the dual problem of TRP can be written in the form DP below.

$$\text{DP:} \quad \text{maximise} \quad \sum_i b_i y_i$$

$$\text{subject to} \quad y_j - y_i \leqslant c_{ij} \qquad \text{for all } i, j \; (i \neq j)$$

$$y_1 = 0.$$

The variables of DP are free because the constraints of TRP are equalities and $y_1 = 0$ reflects the fact that the $i = 1$ equation of (8.1) has been dropped from the constraints of TRP. We could have omitted $y_1$ altogether, but the form DP is generally more convenient. Now, if $t_{ij} = c_{ij} + y_i - y_j$, so that $t_{ij}$ is a slack variable in DP, then

$$-z = \sum_i \sum_{j \neq i} c_{ij} x_{ij} = \sum_i \sum_{j \neq i} t_{ij} x_{ij} - \sum_i y_i \sum_{j \neq i} x_{ij} + \sum_j y_j \sum_{i \neq j} x_{ij}$$

$$= \sum_i \sum_{j \neq i} t_{ij} x_{ij} + \sum_i y_i \left( \sum_{j \neq i} x_{ji} - \sum_{j \neq i} x_{ij} \right),$$

where we have swapped labels $i$ and $j$ in the final sum in the first line to obtain the second line. Hence,

$$z + \sum_i \sum_{j \neq i} t_{ij} x_{ij} = -\sum_i b_i y_i. \tag{8.2}$$

For a given BFS, if we choose the $y_i$ so that $t_{ij} = 0$ whenever $x_{ij}$ is basic, then (8.2) contains only the non-basic variables and we can deduce that $t_{ij}$ is the objective-row coefficient of $x_{ij}$. This means that, if $t_{ij} \geqslant 0$ for all $i, j$ $(i \neq j)$, then the BFS is optimal (the CS conditions) and if not, we make $x_{IJ}$ basic where

$$-t_{IJ} = \max \left\{ -t_{ij} \mid x_{ij} \text{ non-basic} \right\}.$$

To apply this method, we must calculate $y_i$ satisfying

$$y_1 = 0 \text{ and } y_j - y_i = c_{ij} \quad \text{for } x_{ij} \text{ basic.} \tag{8.3}$$

This is easily done by working down the tree. Once all the $y_i$s for the $r$th generation have been determined, we can write them down for the $(r + 1)$th generation using (8.3). If $i$ is in the $(r + 1)$th generation,

$$y_i = y_{P(i)} + c_{P(i),i} \qquad \text{if } x_{P(i),i} \text{ is basic}$$

$$y_i = y_{P(i)} - c_{i,P(i)} \qquad \text{if } x_{i,P(i)} \text{ is basic.}$$

| $c_{ij}$ ($i$ \ $j$) | 1 | 2 | 3 | 4 | 5 | 6 | $b_i$ |
|---|---|---|---|---|---|---|---|
| 1 | – | 5 | 9 | 2 | 4 | 9 | 8 |
| 2 | 1 | – | 2 | 3 | 4 | 8 | –7 |
| 3 | 4 | 5 | – | 6 | 1 | 5 | 3 |
| 4 | 2 | 3 | 6 | – | 8 | 8 | –9 |
| 5 | 8 | 7 | 2 | 7 | – | 1 | 1 |
| 6 | 1 | 4 | 6 | 1 | 2 | – | 4 |

Table 8.1

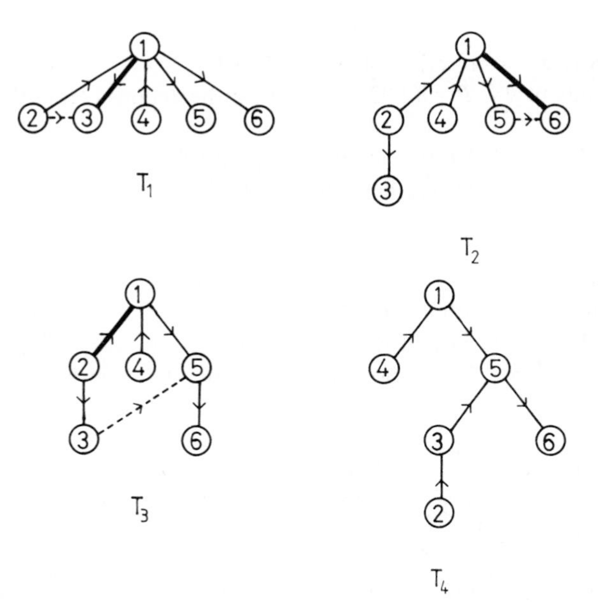


Fig. 8.3 Trees in the solution of P1.

For example, suppose $m = 6$ and the data is given in Table 8.1. For the tree $T_3$ of Fig. 8.3, $y_1 = 0$ and $y_5 = 4$ since $c_{15} = 4$, and $y_2 = -1, y_4 = -2$ since $c_{21} = 1, c_{41} = 2$. This completes the first generation. Then $y_3 = y_2 + c_{23} = -1 + 2 = 1$ and $y_6 = 5$. With these values it is easy to calculate $t_{ij}$ for all non-basic variables, which shows that $t_{25} = -1$, $t_{35} = -2$ and $t_{ij} \geqslant 0$ for all other $i, j$. Thus, we must make $x_{35}$ basic. This will lead to the tree $T_4$ (details below). In $T_4$ we find, working down the tree, $y_1 = 0, y_4 = -2, y_5 = 4, y_3 = 3, y_6 = 5, y_2 = 1$. It is then straightforward to verify that $t_{ij} \geqslant 0$, for all $i, j$, so that the BFS corresponding to $T_4$ is optimal.

We will now give the complete solution of the problem P1 with data specified in Table 8.1 and references to trees $T_1 - T_4$ in Fig. 8.3.

*Iteration 1*

We take as initial BFS: $x_{13} = 3, x_{15} = 1, x_{16} = 4, x_{21} = 7, x_{41} = 9$. This is the initial BFS suggested in Section 8.1 and has cost $9 \times 3 + 4 \times 1 + 9 \times 4 + 1 \times 7 + 2 \times 9 = 92$. It corresponds to the tree $T_1$, which is defined by the parent function $P = (P(2), P(3), \ldots, P(6)) = (1, 1, 1, 1, 1)$.

The dual variables are $y = (y_1, y_2, \ldots, y_6) = (0, -1, 9, -2, 4, 9)$ and the most negative $t_{ij}$ is $t_{23} = -8$. If the edge 2 to 3 is added to $T_4$ (shown dashed in Fig. 8.3), the cycle 2 to 3, 3 to 1, 1 to 2 is created. The reverse variables for this cycle are $x_{13}$ and $x_{21}$. So $x_{13}$ (shown with a heavy line in Fig. 8.3) becomes non-basic, with $\theta = 3$.

*Iteration 2*

The new BFS is $x_{15} = 1, x_{16} = 4, x_{21} = 7 - 3 = 4, x_{23} = 3, x_{41} = 9$, costs 68 and corresponds to $T_2$. $P = (1, 2, 1, 1, 1)$ can be deduced from the rules of Section 8.1.

$y = (0, -1, 1, -2, 4, 9)$ and the most negative $t_{ij}$ is $t_{56} = -4$, giving the cycle 5 to 6, 6 to 1, 1 to 5. So $x_{16}$ becomes non-basic with $\theta = 4$.

*Iteration 3*

The new BFS is $x_{15} = 1 + 4 = 5, x_{21} = 4, x_{23} = 3, x_{41} = 9, x_{56} = 4$, costs 52 and corresponds to $T_3$ with $P = (1, 2, 1, 1, 5)$.

$y = (0, -1, 1, -2, 4, 5)$ and the most negative $t_{ij}$ is $t_{35} = -2$, giving the cycle 3 to 5, 5 to 1, 1 to 2, 2 to 3. So $x_{21}$ becomes non-basic with $\theta = 4$.

*Iteration 4*

The new BFS is $x_{15} = 5 - 4 = 1, x_{23} = 3 + 4 = 7, x_{35} = 4, x_{41} = 9, x_{56} = 4$, costs 44 and corresponds to $T_4$ with $P = (3, 5, 1, 1, 5)$. We have seen above that this solution is optimal.

Following this example through shows that, when solving small examples by hand, it is usually easier to work with the tree rather than the function $P$

defining it. However, it is important to recognise that all the operations required can be performed using $P$ or, more easily, if $G(i)$, the generation of $i$, is available for each $i$ (Exercise 3), although this involves updating $G$ when pivoting.

In the solution above, we calculated $y$ *ab initio* at each iteration. In fact, the calculation of the new $y$ can be simplified by recognising that, in the notation of Section 8.1, if the edge $p$ to $q$ is deleted without adding the edge $I$ to $J$, the tree $T$ is broken up into two smaller **sub-trees.** One of these, which we will call $T^{\mathrm{A}}$, contains node 1. Thus, for $p = 2, q = 9$, the sub-tree $T^{\mathrm{A}}$ in Fig. 8.1(a) is defined by the nodes 1, 9, 4, 5, 7 and 8. It follows from (8.3) that $y_i$ does not change for any $i$ in $T^{\mathrm{A}}$. If $J$ is in $T^{\mathrm{A}}$, then $I$ is in the other sub-tree, $T^{\mathrm{B}}$. Since $y_1 = 0$ does not apply to $T^{\mathrm{B}}$ (because 1 is not a node of $T^{\mathrm{B}}$), if $y_I$ is increased by $\delta$, a solution of (8.3) is obtained by increasing $y_i$, for all nodes of $T^{\mathrm{B}}$, by $\delta$. The increase in $y_I$ is

$$\delta = \text{new}\, y_I - \text{old}\, y_I = (y_J - c_{IJ}) - y_I = -t_{IJ}\ (> 0)$$

and so we increase $y_i$ for all nodes $i$ of $T^{\mathrm{B}}$ by $-t_{IJ}$. By a similar argument, if $I$ is in $T^{\mathrm{A}}$ and $J$ in $T^{\mathrm{B}}$, we decrease $y_i$ for all nodes of $T^{\mathrm{B}}$ by $-t_{IJ}$. This can be verified in Fig. 8.3.

In order to apply this method to large-scale problems, we must find the tree $T^{\mathrm{B}}$ using the pointers available. Although this can be done with $P(.)$ and $G(.)$, it is not particularly efficient and may even be worse than solving (8.3) from scratch. To make it easier to obtain sub-trees, the **children** of any node $i$ (the nodes $j$ for which $P(j) = i$) can be listed. These lists must then be updated when pivoting. More subtle and effective, but much more involved, ways of achieving the same end have been devised.

Assuming non-degeneracy, at each iteration the cost decreases by a positive amount [$-t_{IJ}\, \phi_{pq}$ by (8.2)] and so an optimal BFS is eventually reached. Degeneracy can cause cycling even in TRPs but the possibility of this happening in practice is remote and can be ignored. In Section 4.4, we showed that the pivot can always be chosen to avoid cycling and the procedure described in this section allows us to conclude that if *$b_1, \ldots, b_m$ are all integers, then the optimal solution* (indeed, any BFS) *is integer-valued.* Apart from its intrinsic interest, this result sometimes allows one to formulate certain combinatorial problems, such as the assignment problem discussed in Section 8.4, as TRPs.

A simpler procedure than that of Section 3.6 is available for preventing cycling in TRPs. It rests upon the fact that, if $x_{pq} = 0$ is basic, then $\sum b_i = 0$ where the sum is restricted to the nodes in either of the sub-trees resulting from the removal of the edge $p$ to $q$ from the tree corresponding to the BFS. If we change every $b_i \geqslant 0$ to $b_i + \epsilon$ and any *one* $b_i < 0$ to $b_i - n\epsilon$, where $\epsilon$ is small and $n$ is the number of $b_i \geqslant 0$, then we can never have $\sum b_i = 0$ for any subset of nodes. Consequently, degeneracy cannot occur and, assuming $b_i, \ldots, b_m$ are integers, it is only necessary to round down the optimal solution to the perturbed problem to obtain an optimal solution to the original problem.

## 8.3 TRANSPORTATION PROBLEMS

In some TRPs, many routes between sites will be impossible and so, for many $i$, $j$, $c_{ij}$ will not be specified and we will expect to have $x_{ij} = 0$. Indeed, large problems are usually very sparse (most $x_{ij}$s must be zero). It is easy to cope with this in the TRP framework by setting $c_{ij}$ to some large number $K$ (larger than the sum of the given $c_{ij}$s will do). This will force $x_{ij}$ to be zero in any optimal solution. To capitalise on the computational advantage accruing from missing routes, we never evaluate $t_{ij}$ for any non-basic $x_{ij}$ for which $c_{ij} = K$, since such $x_{ij}$ will never become basic. However, such $x_{ij}$ could be basic in an initial BFS and, although they will eventually leave the basis, it is usually computationally advantageous to start with a BFS in which no basic variable $x_{ij}$ has $c_{ij} = K$. This can often be easily achieved when the problem has sufficient structure as we will now illustrate by turning to transportation problems.

A certain company owns $M$ warehouses and $N$ shops. Warehouse $i$ contains $a_i$ items of a certain product and shop $j$ requires $b_j$ items. If it costs $c_{ij}$ to transport one item from $i$ to $j$, how should the goods be transported in order to satisfy all requirements at minimum cost? If we write $x_{ij}$ for the number of items sent from $i$ to $j$, the problem can be written as an LP. It is called the **Transportation Problem** TP.

$$
\begin{array}{llll}
\text{TP:} & \text{minimise} & \sum_{i,j} c_{ij} x_{ij} & \\
 & \text{subject to} & \sum_{j} x_{ij} = a_i & \text{for } i = 1, \ldots, M \\
 & & \sum_{i} x_{ij} = b_j & \text{for } j = 1, \ldots, N \\
 & & x_{ij} \geqslant 0 & \text{for all } i, j.
\end{array}
$$

For feasibility of TP, we require requirements and availabilities to match: $\sum a_i = \sum b_j$ and then drop the first equation. It should be clear that TP is a special case of TRP with $M + N$ sites, in which we have written $a_i$ instead of $-b_i$ whenever $b_i < 0$ (a warehouse). We also put $c_{ij} = K$ except where $i$ is a warehouse and $j$ a shop, but we have omitted $x_{ij}$ when $c_{ij} = K$, so $K$ does not appear explicitly. The solution procedure for TRPs can obviously be applied to TPs but certain special features arise from the extra structure.

We must first find an initial BFS for TP and at the same time we will generate the parent function defining its corresponding tree. Indeed, the ability to obtain $P(.)$ corresponding to a feasible solution shows that this solution is indeed a BFS, since we explained in Section 8.1 how the canonical equations could be derived from the corresponding tree. The general idea is to select $i$ and $j$ and then make $x_{ij}$ basic and as large as possible without violating the constraints of TP. If

$x_{ij}$ is the first basic variable chosen, this means putting $x_{ij} = \min\{a_i, b_j\}$. Then $a_i$ and $b_j$ must be reduced by this amount, which will leave one of them at zero. The corresponding constraint is made unavailable for further selection and the process repeated until a BFS is obtained. We can refine this process by allowing costs to influence our choice of $i$ and $j$ (small costs of basic variables being desirable), and one such refinement is now described. Others, generally giving a better (smaller cost) initial BFS but at much greater computational cost, have been devised.

We will set out the data in a tableau with rows corresponding to warehouses and columns to shops. Thus, $i$ indexes rows and $j$ indexes columns. A **line** will mean a row or column. When constructing the tree corresponding to a BFS, each line has a corresponding node and we will write R$i$ for the node corresponding to row $i$ and C$j$ for the node corresponding to column $j$. Initially, we will set $\alpha_i = a_i$, $\beta_j = b_j$ for all $i$ and $j$, assume all lines are available and take the current line to be row 1. The general step is to find the smallest available cost in the current line, say $c_{ij}$, and put $x_{ij} = \min\{\alpha_i, \beta_j\}$. Then we set

(i) $P(\mathrm{C}j) = \mathrm{R}i$, if the current line is row $i$,
(ii) $P(\mathrm{R}i) = \mathrm{C}j$, if the current line is column $j$.

(a) If $x_{ij} = \alpha_i$, row $i$ is no longer available, set the current line to column $j$ and change $\beta_j$ to $\beta_j - \alpha_i$;

(b) if $x_{ij} > \alpha_i$, column $j$ is no longer available, set the current line to row $i$ and change $\alpha_i$ to $\alpha_i - \beta_j$.

This step is repeated until $M + N - 1$ basic variables have been chosen.

Although complicated to describe, this method is intuitive and easy to apply, and we will illustrate it by considering problem P2 whose data is displayed in tableau T1 below. The entry in row $i$ and column $j$ of T1 is $c_{ij}$. Since $M = 3$, $N = 4$, there are 6 basic variables to be found.

| T1 | 1 | 2 | 3 | 4 | $a_i$ |
|---|---|---|---|---|---|
| 1 | 9 | 4 | 7 | 8 | 9 |
| 2 | 2 | 5 | 3 | 6 | 7 |
| 3 | 8 | 7 | 2 | 7 | 5 |
| $b_j$ | 8 | 3 | 2 | 8 | |

| T2 | 4 | 4 | 7 | 8 |
|---|---|---|---|---|
| 0 | 5 | (3) | (2) | (4) |
| −2 | (3) | 3 | −2 | (4) |
| 4 | (5) | −1 | −9 | −5 |

1. Current line is row 1. Least cost is $c_{12} = 4$. Put $x_{12} = \min\{3, 9\} = 3$, $P(\mathrm{C}2) = \mathrm{R}1$; change $\alpha_1$ to 6.
2. Current line is still row 1. $c_{13}$ least. Put $x_{13} = 2, P(\mathrm{C}3) = \mathrm{R}1$; change $\alpha_1$ to 4.

3. Current line: row 1, $c_{14}$ least. Put $x_{14} = 4, P(\text{C4}) = \text{R1}$; change $\beta_4$ to 4.
4. Current line: column 4, $c_{24}$ least. Put $x_{24} = 4, P(\text{R2}) = \text{C4}$; change $\alpha_2$ to 3.
5. Current line: row 2, $c_{21}$ least. Put $x_{21} = 3, P(\text{C1}) = \text{R2}$; change $\beta_1$ to 5.
6. Current line: column 1. Only row 3 still available $x_{31} = 5, P(\text{R3}) = \text{C1}$.

The initial BFS produced by this method can also be used as an initial BFS for the TRP by identifying $i$ with $b_i < 0$ as warehouses and $i$ with $b_i \geqslant 0$ as shops and allowing the initial BFS to contain only $x_{ij}$, where $i$ is a warehouse and $j$ a shop. Applied to problem P1 of Section 8.2, it gives $x_{21} = 7, x_{41} = 1, x_{43} = 3$, $x_{45} = 1$ and $x_{46} = 4$, which has a cost of 67 compared with 92 for the initial BFS given in that section.

The dual problem to TP has dual variables $y_i$ and $z_j$, and constraints

$$y_i + z_j \leqslant c_{ij} \qquad \text{for all } i, j \ (i \neq j) \text{ and } y_1 = 0.$$

So, for each basic $x_{ij}$ we have $y_i + z_j = c_{ij}$, and $y_i$ and $z_j$ can be determined by working down the tree as in TRP. Then we calculate

$$t_{ij} = c_{ij} - y_i - z_j$$

and proceed exactly as for the TRP. The solution can be set out neatly in tabular form as in T2, T3, T4 where the circled entries are the values of basic variables and the remaining entries display $t_{ij}$ (which is zero for basic variables). The BFS given in T2 is the initial solution obtained above. Variables $y_i$ and $z_j$ are also shown in T2–T4 (in the first column and top row, respectively).

| T3 | 4 | 4 | −2 | 8 |
|---|---|---|---|---|
| 0 | 5 | (3) | 9 | (6) |
| −2 | (5) | 3 | 7 | (2) |
| 4 | (3) | −1 | (2) | −5 |

| T4 | 9 | 4 | 3 | 8 |
|---|---|---|---|---|
| 0 | 0 | (3) | 4 | (6) |
| −7 | (7) | 8 | 7 | 5 |
| −1 | (1) | 4 | (2) | (2) |

The trees corresponding to T2–T4 are $T_2$, $T_3$, $T_4$ of Fig. 8.4. It is clear that any basic variable $x_{ij}$ corresponds to an edge R$i$ to C$j$ so there is no need to put arrows on the edges. Indeed, if the edge R$I$ to C$J$ is added, the path from C$J$ to R$I$ corresponds to an alternating sequence of reverse and forward variables, starting with a reverse variable. This can be verified in $T_2$ and $T_3$.

In some circumstances we may wish to solve **unbalanced** problems ($\sum a_i \neq \sum b_j$). For example, suppose $\sum a_i < \sum b_j$, so that some shops will receive less than their requirements. Then the constraints can be written

$$\sum_j x_{ij} = a_i \qquad \text{for } i = 1, \ldots, m$$

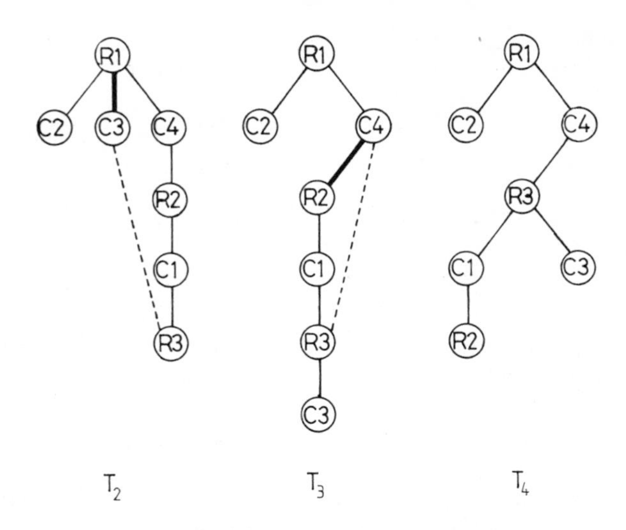


Fig. 8.4 Trees in the solution of P2.

$$\sum_i x_{ij} \leqslant b_j \qquad \text{for } j = 1, \ldots, n.$$

If we now impose a penalty $p_j$ ($\geqslant 0$) per unit shortfall at shop $j$, the LP formulation becomes

$$\begin{aligned} &\text{minimise} && \sum_{i,j} c_{ij} x_{ij} + \sum_j p_j s_j && \\ &\text{subject to} && \sum_j x_{ij} = a_i && \text{for } i = 1, \ldots, m \\ & && \sum_i x_{ij} + s_j = b_j && \text{for } j = 1, \ldots, n \\ & && x_{ij}, s_j \geqslant 0 && \text{for all } i, j \end{aligned}$$

which is just a TP with $m + 1$ warehouses, if we rewrite $s_j$ as $x_{m+1,j}$ and add the constraint $\sum x_{m+1,j} = \sum b_j - \sum a_i$. A similar trick can be used if $\sum a_i > \sum b_j$.

## 8.4 THE ASSIGNMENT PROBLEM

Suppose we have $M$ individuals, each of whom can perform any of $M$ tasks. The cost of assigning individual $i$ to task $j$ is $c_{ij}$ and the **Assignment Problem**, AP, seeks to assign each individual to a different task at minimum total cost. If we write $x_{ij} = 1$ if individual $i$ is assigned to task $j$ and $x_{ij} = 0$ otherwise, the problem can be written:

$$\text{AP: minimise} \quad \sum_{i,j} c_{ij} x_{ij}$$

$$\begin{aligned} \text{subject to} \quad & \sum_j x_{ij} = 1 && \text{for } i = 1, \ldots, M \\ & \sum_i x_{ij} = 1 && \text{for } j = 1, \ldots, M \\ & x_{ij} = 0, \text{ or } 1 && \text{for all } i, j. \end{aligned}$$

The first $M$ constraints express the fact that each individual undertakes exactly one task and the second $M$ constraints say that each task should be undertaken by one individual.

If we were to relax the requirement $x_{ij} = 0$ or 1 to $x_{ij} \geqslant 0$, we would have a TP, but any TP has an optimal solution in integers and, in view of the constraints, this means $x_{ij} = 0$ or 1. So the AP can be viewed as a TP. However, it is a highly degenerate TP which, at worst, can mean that the problem cycles and always brings the danger that a very large number of iterations may be required. Nevertheless, solving an AP as a TP works very well, if we restrict the solution to a special class of BFSs.

Any BFS of AP must have one 1 in each row and column. This gives $M$ 1s, but a basis must contain $2M - 1$ variables, so we need $(M - 1)$ 0s among the basic variables. We will describe a basis as **alternating** if the first row contains one basic variable (which must be a 1) and the remaining rows contain two basic variables (a 1 and a 0). If $x_{IJ}$ enters the basis, we will typically have a choice of variable to leave the basis. To ensure that the new basis is alternating, we will make a basic variable in row $I$ non-basic. Clearly, exactly one of these basic variables, say $x_{IK}$, will correspond to an edge of the cycle created by adding the edge $I$ to $J$ to the tree and it can leave the basis provided it becomes 0 in the new BFS. This will certainly be true if $x_{IK}$ is 0 in the current BFS, because it is a reverse variable (why?) so $\theta = 0$ and its value does not change (a degenerate pivot). If $x_{IK}$ is 1, we must show that $\theta = 1$, so that $x_{IK}$ decreases to 0.

To see that this is true, we note that any path in a tree corresponding to a BFS of an AP consists of an alternating sequence of row and column vertices, and R$i$ and C$j$ adjacent in this sequence means $x_{ij}$ is basic. Now suppose that the path from C$J$ to R$I$ ends: $\ldots \to$ C$j \to$ R$i \to$ C$K \to$ R$I$, and $\phi_{IK} = 1$. Then $\phi_{iK} = 0$, for otherwise there would be two 1s in the $K$th column. Hence $\phi_{ij} = 1$, since there is no other basic variable in row $i$. (We cannot have $i = 1$, though $I = 1$ is possible.) By continuing in this fashion, we see that the path from C$J$ to R$I$ corresponds to an alternating sequence of reverse variables with value 1 and forward variables with value 0. This means that $\theta = 1$.

To start the method, we must find an initial alternating basis. This is done by first obtaining an initial assignment as follows. We first look for the smallest $c_{ij}$, say $c_{tu} = \min_{i,j} c_{ij}$, and set $x_{tu} = 1$. Then we declare row $t$ and column $u$ unavailable and search for the smallest $c_{ij}$ in the remaining available rows and columns, say $c_{vw}$, and set $x_{vw} = 1$, declaring row $v$ and column $w$ unavailable. This process is repeated until there is an $x_{ij} = 1$ in every row $i$. Applied to P3, which has costs displayed in tableau T5, it gives, in order, $x_{23} = 1, x_{44} = 1, x_{12} = 1, x_{31} = 1$.

To complete an alternating BFS, we must find a further basic variable (with value 0) for each row $i \geqslant 2$. To do this, we choose $x_{ij}$ basic, where $x_{rj} = 1$ for some $r < i$ and

$$c_{ij} = \min \{c_{ik} \mid x_{rk} = 1 \ \text{ for some } r < i\}.$$

Expressed verbally, we must choose a variable with a 1 above it in the same column. From those available in row $i$ ($\geqslant 2$), the one with smallest cost is selected. Applied to P3, this gives the BFS displayed in T6. The parent function of the corresponding tree is given by putting $P(\mathrm{R}i) = \mathrm{C}j$ and $P(\mathrm{C}l) = \mathrm{R}i$, where $x_{il} = 1$. For T6, the tree is shown as $T_6$ in Fig. 8.5. Note that $x_{ij} = 1\,[0]$, if $P(\mathrm{C}j) = \mathrm{R}i\,[P(\mathrm{R}i) = \mathrm{C}j]$.

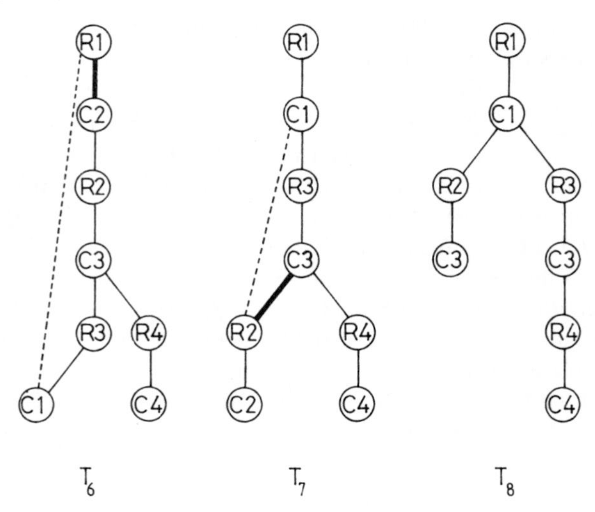


Fig. 8.5 Trees in the solution of P3.

The complete solution is shown in tableaux T6–T8; the corresponding trees are $T_6 - T_8$ of Fig. 8.5. Note that, in T6, we make $x_{11}$ basic (we could also have chosen $x_{21}$), but $x_{12}$ *must* leave the basis (not $x_{23}$ or $x_{31}$) in order to maintain an alternating BFS. Similarly, in T7, $x_{23}$ (not $x_{31}$) leaves the basis.

This method cannot cycle. To see this, we first observe that, if $\theta = 1$ the cost decreases (by $-t_{IJ}$) so the preceding BFS cannot recur. When $\theta = 0$, which

| T5 | 1 | 2 | 3 | 4 |
|---|---|---|---|---|
| 1 | 7 | 6 | 6 | 6 |
| 2 | 3 | 2 | 1 | 3 |
| 3 | 8 | 7 | 2 | 7 |
| 4 | 9 | 7 | 2 | 5 |

| T6 | 11 | 6 | 5 | 8 |
|---|---|---|---|---|
| 0 | –4 | ① | 1 | –2 |
| –4 | –4 | ⓪ | ① | –1 |
| –3 | ① | 4 | ⓪ | 2 |
| –3 | 1 | 4 | ⓪ | ① |

| T7 | 7 | 2 | 1 | 4 |
|---|---|---|---|---|
| 0 | ① | 4 | 5 | 2 |
| 0 | –4 | ① | ⓪ | –1 |
| 1 | ⓪ | 4 | ① | 2 |
| 1 | 1 | 4 | ⓪ | ① |

| T8 | 7 | 6 | 1 | 4 |
|---|---|---|---|---|
| 0 | ① | 0 | 5 | 2 |
| –4 | ⓪ | ① | 4 | 3 |
| 1 | ⓪ | 0 | ① | 2 |
| 1 | 1 | 0 | ⓪ | ① |

means $\phi_{IK} = 0$, as we saw above, we shall show that $\sum z_i - \sum y_i$ increases, so the preceding BFS cannot recur in this case, either, since $y_i$ and $z_j$ are uniquely determined by the basis (and $y_1 = 0$).

To see that $\sum z_i - \sum y_i$ increases, we first observe that this sum is just the sum of all the dual variables of the AP viewed as a TRP. (We changed the sign of the dual variables corresponding to rows when going from TRP to TP.) To prove that this sum increases we have only to show that, in the notation of Section 8.2, C$J$ is in $T^{\mathrm{A}}$. (This actually shows that some $z_i, -y_i$ increase by $-t_{IJ}$ and none decreases.) But, by a similar argument to that used above, we can see that the path from R1 to R$I$ corresponds to an alternating sequence of forward variables with value 1 and reverse variables with value 0. Hence, if C$k$ is the penultimate node on this path, $x_{Ik}$ is basic and $\phi_{Ik} = 0$. But there is only one variable, $x_{IK}$, with these properties in row $I$, so $k = K$ and removing edge R$I$ to C$K$ breaks the path from R1 to R$I$. This puts R$I$ in $T^{\mathrm{B}}$ and C$J$ in $T^{\mathrm{A}}$.

## 8.5 CAPACITY CONSTRAINTS

In this section, we will impose the extra constraint

$$0 \leqslant x_{ij} \leqslant U_{ij} \quad \text{or} \quad x_{ij} + \sigma_{ij} = U_{ij}, \quad x_{ij} \geqslant 0, \quad \sigma_{ij} \geqslant 0$$

on all, or some, of the variables of TRP. We will see how such upper-bound constraints can be handled implicitly as in Chapter 7. As in that chapter, we allow non-basic variables to be $x_{ij}$ or $\sigma_{ij}$. If $y_i$ $(i \geqslant 1)$ satisfies (8.3), then $t_{ij}$ is the coefficient of non-basic $x_{ij}$ so that $-t_{ij}$ is the coefficient of $\sigma_{ij}$. We calculate

(i) $\quad -t_{IJ} = \max\{-t_{ij} \mid x_{ij} \text{ non-basic}\}$,
(ii) $\quad t_{KL} = \max\{t_{ij} \mid \sigma_{ij} \text{ non-basic, i.e. } x_{ij} = U_{ij}\}$,

and, if these are both non-positive, the current BFS is optimal.

If $t_{IJ} < 0$ and $-t_{IJ} \geqslant t_{KL}$, we make $x_{IJ}$ basic. This is done as in Section 8.2, by finding the cycle created when the edge $I$ to $J$ is added to the tree corresponding to the *explicit* basic variables of the current BFS. We saw in Section 8.1 that the coefficient of $x_{IJ}$ in the equation of the canonical form in which $x_{ij}$ is basic is $\delta_{ij}$, where $\delta_{ij} = 1$ for reverse and $-1$ for forward variables. This means that the coefficient of $x_{IJ}$ in the equation in which $\sigma_{ij}$ is basic is 1, if $x_{ij}$ is a forward basic variable, and the RHS of this equation is then $U_{ij} - \phi_{ij}$. Also, the coefficient of $x_{IJ}$ is 1 in the equation in which $\sigma_{IJ}$ is basic and the RHS is $U_{IJ}$. (For further details, see Chapter 7.) Thus, we define $\theta$ to be the minimum of

(a) $\quad \phi_{ij}$ for reverse variables $x_{ij}$,
(b) $\quad U_{ij} - \phi_{ij}$ for forward variables $x_{ij}$,
(c) $\quad U_{IJ}$.

The new variable values become $\phi_{ij} + \theta$ for forward $x_{ij}$, $\phi_{ij} - \theta$ for reverse $x_{ij}$ and remain at $\phi_{ij}$ for the remaining variables. If the minimum occurs under case (a) for $\phi_{pq}$, we make $x_{pq}$ non-basic and, if the minimum occurs under case (b) for $U_{pq} - \phi_{pq}$, we make $\sigma_{pq}$ non-basic, so $x_{pq}$ becomes $U_{pq}$. In either case, we add edge $I$ to $J$ to the tree and delete $p$ to $q$. The calculation of the new parent function is done as in Section 8.1, except that for case (b), we reverse the roles of $p$ and $q$. If the minimum occurs under case (c), we make $\sigma_{IJ}$ non-basic (instead of $x_{IJ}$) and do not change the tree.

These rules can also be deduced by imagining that we try to send extra goods 'round the cycle' by increasing the flow of goods in edges corresponding to forward variables and reducing it in edges corresponding to reverse variables. The requirement that no variable should exceed its upper bound or become negative leads to the results above.

If $t_{KL} > 0$ and $t_{KL} > -t_{IJ}$ in (i) and (ii) above, we make $\sigma_{KL}$ basic. This is done by adding an edge $L$ to $K$, thus creating a cycle. The direction of the cycle is chosen so that the edge $L$ to $K$ is traversed from $L$ to $K$. Then we define $\theta$ as above, but with $U_{KL}$ replacing $U_{IJ}$ in (c). The rest of the procedure is the same as before (noting that $x_{KL}$ *decreases* by $\theta$) except that, if the minimum occurs in case (c), we make $x_{KL}$ non-basic (instead of $\sigma_{KL}$).

An extra difficulty arises in problems with upper bounds in that the initial BFS may be infeasible because some $x_{ij}$ exceed their upper bounds. In Chapter 3, we saw that, for general LPs, a two-phase method, maximising $z^{\mathrm{I}}$ in Phase I, could be used to tackle such problems. The same approach can be adopted for TRPs since Phase I can itself be regarded as a TRP. However, it proves convenient to combine the two phases by replacing the cost $c_{ij}$ by $c_{ij} + K$, if $x_{ij}$ is a basic variable with $\phi_{ij} > U_{ij}$, where $K$ is a large number. Provided $K$ is large enough and there is a feasible solution, no optimal solution, for any set of costs modified in this way, contains basic variables exceeding their upper bound. Thus, we will eventually achieve an optimal feasible solution. We will not specify $K$ explicitly, but will leave it as a parameter. Consequently, any $t_{ij}$ will be of the form $\alpha_{ij} + \beta_{ij}K$ and to compare $t_{ij}$, we first compare $\beta_{ij}$ and, only if these are the same, compare $\alpha_{ij}$. This means that we are effectively using the two-phase method, except that, where a choice of variable to enter the basis occurs in Phase I (and such occurrences are very common), the choice is resolved by the true objective function.

While some basic variables remain infeasible, the rule for selection of the variable to leave the basis needs modification. In fact, all that is required is that, when calculating $\theta$ as above, we restrict case (b) to feasible forward variables ($\phi_{ij} \leqslant U_{ij}$). This must still yield a finite minimum, for otherwise we would have a solution ($x_{ij} = \phi_{ij} + \theta\,\delta_{ij}$) satisfying (8.1) and $x_{ij} \geqslant 0$ for all $\theta \geqslant 0$ and with cost decreasing ($z$ increasing) in $\theta$. By choosing $\theta$ large enough, this would result in a negative cost, which is impossible, since $c_{ij} \geqslant 0$ for all $i, j$.

This modification can also be derived from the modified PRS rule of Section 3.3

We will now solve problem P1 of Section 8.2 again, but with the addition of upper bounds $U_{23} = 2$, $U_{25} = 4$, $U_{41} = 6$, $U_{43} = 2$, $U_{56} = 1$, $U_{45} = 1$. The remaining variables are unrestricted. The corresponding trees are shown in Fig. 8.6 but it should be remembered that non-basic variables can also have a positive $x_{ij}$ ($= U_{ij}$). We will use the same initial BFS as in Section 8.2, but the method of constructing an initial BFS described in Section 8.3 can be modified to produce a good (low cost) feasible, or nearly feasible (few infeasible variables), solution. (See Exercise 10.)

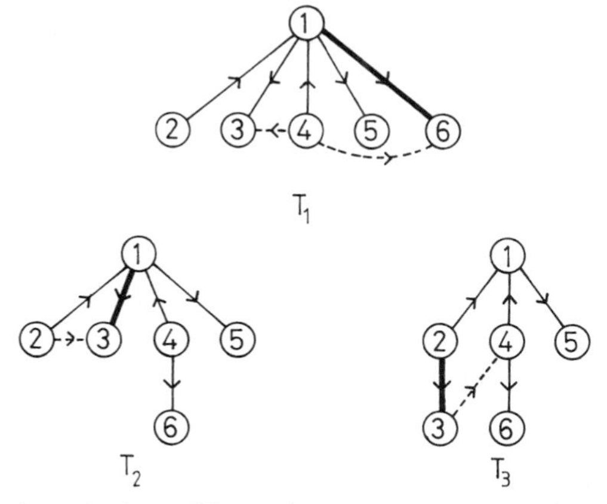


Fig. 8.6 Trees in the solution of P1 with bounded variables (*cont. overleaf*)

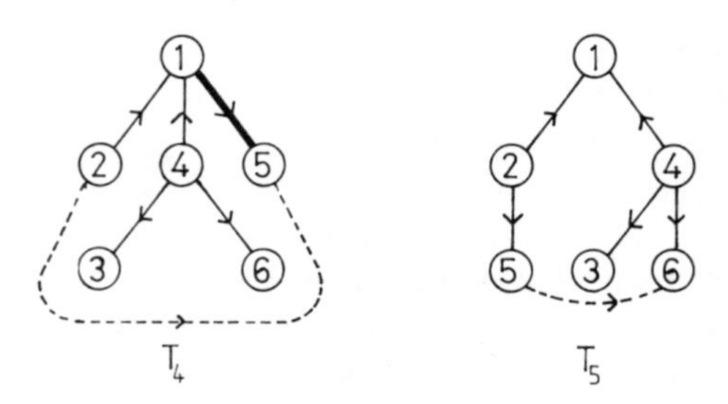


Fig. 8.6 (cont.)

*Iteration 1*

Initial BFS: $x_{13} = 3, x_{15} = 1, x_{16} = 4, x_{21} = 7, x_{41} = 9$, corresponds to $T_1$ and has $P = (1, 1, 1, 1, 1)$. $x_{41} = 9 > U_{41}$; put $c_{41} = K + 2$ *pro tem.* $y = (0, -1, 9, -K-2, 4, 9)$ so $t_{42} = -K+2, t_{43} = -K-5, t_{45} = -K+2, t_{46} = -K-3$ and no other $t_{ij}$ involves $K$. Hence $t_{IJ} = t_{43} = -K-5$ and no $\sigma_{ij}$ is non-basic, so we make $x_{43}$ basic, giving the cycle 4 to 3, 3 to 1, 1 to 4. $\theta$ is the minimum of (a): 3 ($x_{13}$), (b): $\infty$ (there are no forward variables), (c): 2. So $\theta = 2$, $\sigma_{43}$ becomes non-basic ($x_{43}$ becomes 2), but the tree does not change.

*Iteration 2*

The new BFS is $x_{13} = 1, x_{15} = 1, x_{16} = 4, x_{21} = 7, x_{41} = 7, (x_{43} = 2)$ $P, y$ and $t_{ij}$ are all the same as iteration 1. (i) $-t_{IJ} = -t_{46} = K+3$, (ii) $t_{KL} = t_{43} = -K-8$, so we make $x_{46}$ basic, giving the cycle 4 to 6, 6 to 1, 1 to 4. $\theta$ is the minimum of (a): 4 ($x_{16}$), (b): $\infty$, (c): 5. So $x_{16}$ becomes non-basic with $\theta = 4$.

*Iteration 3*

The new BFS: $x_{13} = 1, x_{15} = 1, x_{21} = 7, x_{41} = 3, x_{46} = 4, (x_{43} = 2)$ is feasible (drop $K$), and corresponds to $T_2$ with $P = (1, 1, 1, 1, 4)$. $y = (0, -1, 9, -2, 4, 6)$ and (i) $-t_{IJ} = -t_{23} = 8$, (ii) $t_{KL} = t_{43} = -5$, so we make $x_{23}$ basic, giving the cycle 2 to 3, 3 to 1, 1 to 2. $\theta$ is the minimum of (a): 1 ($x_{13}$), (b): $\infty$, (c): 2. So $x_{13}$ becomes non-basic with $\theta = 1$.

*Iteration 4*

The new BFS is $x_{15} = 1, x_{21} = 6, x_{23} = 1, x_{41} = 3, x_{46} = 4, (x_{43} = 2)$, and corresponds to $T_3$ with $P = (1, 2, 1, 1, 4)$ $y = (0, -1, 1, -2, 4, 6)$ and (i) $-t_{IJ} = -t_{35} = 2$, (ii) $t_{KL} = t_{43} = 3$, so we make $\sigma_{43}$ basic (reduce $x_{43}$) by adding an edge directed from 3 to 4, giving the cycle 3 to 4, 4 to 1, 1 to 2, 2 to 3. $\theta$ is the minimum of (a): 6 ($x_{21}$), (b): $1 = 2 - 1 = U_{23} - x_{23} = \sigma_{23}$, (c): 2. So $\sigma_{23}$ is made non-basic (reduce $x_{23}$) with $\theta = 1$.

*Iteration 5*

The new BFS is $x_{15} = 1, x_{21} = 5, x_{41} = 4, x_{43} = 1, x_{46} = 4, (x_{23} = 2)$, and corresponds to $T_4$ with $P = (1, 4, 1, 1, 4)$. $y = (0, -1, 4, -2, 4, 6)$. Make $x_{25}$ basic, $x_{15}$ non-basic, with $\theta = 1$.

*Iteration 6*

The new BFS is $x_{21} = 4, x_{15} = 1, x_{41} = 4, x_{43} = 1, x_{46} = 4, (x_{23} = 2)$, and corresponds to $T_5$ with $P = (1, 4, 1, 2, 4)$. $y = (0, -1, 4, -2, 3, 6)$. Make $\sigma_{56}$ basic with $\theta = 1$ ($x_{56}$ becomes 1). The tree does not change.

*Iteration 7*

The new BFS is $x_{21} = 3, x_{25} = 2, x_{41} = 5, x_{43} = 1, x_{46} = 3, (x_{23} = 2, x_{56} = 1)$. $P$, $y$, $t_{ij}$ are all the same as iteration 6. $t_{23} = -3$, $t_{56} = -2$ and all other $t_{ij} \geqslant 0$. So current BFS is optimal.

## EXERCISES

1.
(i) Solve the following TRP.

| $i$ \ $c_{ij}$ | $j$ = 1 | 2 | 3 | 4 | 5 | 6 | $b_i$ |
|---|---|---|---|---|---|---|---|
| 1 | – | 1 | 5 | 2 | 4 | 3 | 8 |
| 2 | 6 | – | 4 | 9 | 8 | 5 | −1 |
| 3 | 5 | 8 | – | 7 | 6 | 2 | 5 |
| 4 | 3 | 6 | 2 | – | 4 | 5 | −1 |
| 5 | 7 | 2 | 3 | 2 | – | 9 | −2 |
| 6 | 4 | 8 | 5 | 4 | 3 | – | −9 |

(ii) Solve the same problem when $b_1$ is changed to 11 and $b_5$ to $-5$.

N.B. In both cases use an initial BFS with $P(i) = 1$ for $i \geqslant 2$.

2. Use the method suggested in Section 8.2 for preventing degeneracy to solve Exercise 1(ii).

3. Explain how the use of $G$ can accelerate the computation of the path from $J$ to $I$ in the tree corresponding to a BFS when $x_{IJ}$ enters the basis.

4.
(i) Solve the following TP.

| $c_{ij}$ | 1 | 2 | 3 | 4 | 5 | $a_i$ |
|---|---|---|---|---|---|---|
| 1 | 2 | 1 | 7 | 4 | 3 | 5 |
| 2 | 2 | 1 | 4 | 5 | 3 | 9 |
| 3 | 5 | 2 | 9 | 9 | 1 | 9 |
| $b_j$ | 5 | 2 | 6 | 7 | 3 | |

(ii) Solve the same problem when $a_3$ is changed to 13, $b_2$ to 5 and $b_5$ to 4.
(iii) Solve the unbalanced problem obtained by increasing $a_1$ to 9.

5. Solve Exercise 4(i) when $a_1$ and $b_5$ are both increased by 1.

6. Using the initial BFS of a TP to start the method, solve Exercise 1(i).

7. A manufacturer produces engines over a period of $N$ months. In month $i$ he can produce up to $a_i$ engines and he has to supply $b_i$ engines. He can store engines produced in one month to supply demand in a future month, but storage costs $h$ per month in store plus a fixed cost of $A$ to cover administrative costs, per engine. Explain how the problem of minimising costs can be formulated as a TP. What conditions ensure feasibility?

8. Solve the AP:

| $i$ \ $j$ | 1 | 2 | 3 | 4 | 5 |
|---|---|---|---|---|---|
| 1 | 4 | 6 | 5 | 1 | 2 |
| 2 | 7 | 9 | 9 | 6 | 4 |
| 3 | 5 | 8 | 5 | 5 | 1 |
| 4 | 1 | 3 | 3 | 2 | 1 |
| 5 | 6 | 8 | 7 | 6 | 2 |

9. Solve Exercise 1(i) with capacity constraints $U_{12} = 1$, $U_{14} = 4$, $U_{36} = 2$, $U_{41} = 3$, $U_{43} = 2$, $U_{53} = 1$, $U_{61} = 3$.

10. Suggest how the method of Section 8.3 for finding an initial BFS of a TP may be modified to find a feasible or nearly feasible initial solution when capacity constraints are present. Apply your method to Exercise 4(i) when $U_{ij} = 3$ for all $i, j$ $(i \neq j)$.

CHAPTER 9

# Problems with Multiple Objectives – Efficiency

## 9.1 INTRODUCTION AND WEIGHTING FACTORS

Consider a manufacturer who can produce three products *A, B* and *C*. He has stocks of 250 units of input I and 210 units of input II. Each unit of *A* has 7 units of I and 6 units of II, *B* uses 5 of I and 9 of II and *C* uses 6 of I and 5 of II. If $x_1$, $x_2$, $x_3$ are, respectively, the amounts of *A, B* and *C* produced, the manufacturer is faced with the constraints

$$7x_1 + 5x_2 + 6x_3 \leqslant 250$$

$$6x_1 + 9x_2 + 5x_3 \leqslant 210$$

$$x_1, x_2, x_3 \geqslant 0.$$

The selling prices of *A, B* and *C* are 1, 0.8 and 0.9 respectively so he would like to maximise his total revenue

$$x_1 + 0.8x_2 + 0.9x_3$$

subject to these contraints. However, he also feels that total sales are important, so he would also like to maximise

$$x_1 + x_2 + x_3 .$$

The manufacturer also believes that there is a useful market to be exploited overseas. He would therefore also like to maximise exports. If 40 per cent of *A*, 60 per cent of *B* and 20 per cent of *C* produced are exported, this involves maximising

$$0.4x_1 + 0.6x_2 + 0.2x_3 .$$

These three objectives reflect aspects of the production problem and, ideally, the manufacturer would like a feasible solution maximising them all. Unfortunately such a solution is rarely available and in this and the next chapter we shall examine ways of 'solving' such problems, which are typically grouped together under the title **multiple-objective** or **multiple-criterion** problems.

It is worth noting that multiple objectives also arise when several individuals have to make a joint decision, the objective functions modelling the aspirations of each individual. For example, in the production problem above, the three objectives may be those of the managing director, the sales director and the export director. We shall examine an extreme case of two decision-makers in Section 10.4.

In general, we will examine problems having $p$ linear objective functions, $\sum_j c_{kj} x_j$ for $k = 1, \ldots, p$, which we wish to maximise subject to linear constraints. Typically, these objectives will conflict in that there is no feasible solution which simultaneously maximises them all. In such a case, some form of conflict resolution must be adopted to arrive at a solution. We will discuss some methods of achieving this in Chapter 10. However, one of the simplest and most frequently used approaches will also play an important role in our development of efficiency and so we will introduce it here.

The idea is to multiply each objective by a weighting factor and then add the weighted objectives. For example, if we choose weighting factors of 10 for revenue, 1 for sales and 5 for exports in the example described above, the sum of the weighted objectives is $13x_1 + 12x_2 + 11x_3$ and this is to be maximised subject to the constraints given above. The optimal solution is $(x_1, x_2, x_3) = (10, 0, 30)$. The size of the weighting factor can be chosen to reflect the importance we attach to an objective, but the use of this linear form implicitly assumes that the weights should be independent of the solution and does not permit one to model, for example, the proposal that profits should be given higher priority relative to exports when profits are low than when they are high.

In general, if $w_k$ is the weight ascribed to the $k$th objective and the constraints are taken to be inequalities, we are faced with solving the problem

$$\begin{array}{lll} \text{LP}(w_1, \ldots, w_p): & \text{maximise} & \sum_j \sum_k (w_k c_{kj}) x_j \\ & \text{subject to} & \sum_j a_{ij} x_j \leqslant b_i \quad \text{for all } i \\ & & x_j \geqslant 0 \quad \text{for all } j. \end{array}$$

In Section 9.3 we will investigate the relationship between $\text{LP}(w_1, \ldots, w_p)$ and efficient solutions, introduced in the next section.

## 9.2 EFFICIENT SOLUTIONS

We have assumed that all objective functions are 'desirable' in that we prefer greater to lesser objective function values. Consequently, if a feasible solution $(x_1, \ldots, x_n)$ **dominates** $(y_1, \ldots, y_n)$ in the sense that

$$\sum_j c_{kj} x_j \geqslant \sum_j c_{kj} y_j \qquad \text{for all } k \tag{9.1}$$

and

$$\sum_j c_{kj} x_j > \sum_j c_{kj} y_j \qquad \text{for at least one } k, \tag{8.2}$$

then we would prefer $(x_1, \ldots, x_n)$ to $(y_1, \ldots, y_n)$. The strict inequality (9.2) is used to exclude the possibility of equality in (9.1) for every $k$.

For example, with the objectives

$$x_1 - x_2 + x_3$$

and

$$2x_1 + x_2 - 3x_3$$

the points A: (2, 1, 1), B: (6, 3, 2), C: (6, 6, 3), D: (4, 1, 0), E: (6, 0, 2) have objective function values A: (2, 2), B: (5, 9), C: (3, 9); D: (3, 9), E: (8, 6), so that A is dominated by B, B dominates C, but C dominates neither of D or E.

We will say that a feasible solution is **efficient** if it is dominated by no other feasible solution. Synonyms for 'efficient' include 'admissible', 'Pareto optimal' and 'non-dominated'. It is a consequence of the definition that it is impossible to increase any objective function at an efficient solution without decreasing another. In the multi-person case this says that no individual can be made better off without making someone else worse off. Concentrating on efficiency means that we do not need to select a method of conflict resolution (or at least may postpone the selection). The compensating disadvantage is that when conflict exists there will be more than one efficient solution, but at least we would expect to have excluded those non-preferred solutions which could never be optimal.

When we are studying a multiple-objective problem with a view to calculating efficient points, we will speak of a **vector maximisation problem** (VMP) and display the data with the objective functions above the constraints as in the following example, P1.

$$
\begin{array}{llrl}
\text{P1:} & V\max & x_1 - x_2 + x_3 & (= z_1) \\
 & & -x_1 - 2x_2 + x_3 & (= z_2) \\
 & & 2x_1 + x_2 + x_3 & (= z_3) \\
 & \text{subject to} & x_1 + 3x_2 + x_3 \leqslant 5 & \\
 & & 2x_1 + x_2 + 2x_3 \leqslant 4 & \\
 & & x_1, x_2, x_3 \geqslant 0 &
\end{array}
$$

or, more generally,

$$
\begin{array}{lll}
V\max & \sum_j c_{kj}x_j & \text{for } k = 1, \ldots, p \\
\text{subject to} & \sum_j a_{ij}x_j \leqslant b_i & \text{for all } i \\
& x_j \geqslant 0 & \text{for all } j
\end{array}
$$

when the constraints are written in inequality form.

We would like to be able to test whether a feasible solution $(\hat{x}_1, \ldots, \hat{x}_n)$ is efficient. As it stands, the strict inequality (9.2) is rather hard to handle, but (9.1) and (9.2) together are equivalent to (9.1) together with the inequality:

$$\sum_k \sum_j c_{kj}x_j > \sum_k \sum_j c_{kj}y_j. \tag{9.3}$$

(These expressions are the objective function values of LP(1, 1, . . . , 1).) Consequently, if we define

$$
\begin{array}{llll}
\text{LP}^*(\hat{x}_1, \ldots, \hat{x}_n)\text{: maximise} & \sum_k \sum_j c_{kj}x_j & & \\
\text{subject to} & \sum_j c_{kj}x_j \geqslant \sum_j c_{kj}\hat{x}_j & \text{for all } k \\
& \sum_j a_{ij}x_j \leqslant b_i & \text{for all } i \\
& x_i \geqslant 0 &
\end{array}
$$

then, either $(\hat{x}_1, \ldots, \hat{x}_n)$ is optimal, or any optimal solution dominates $(\hat{x}_1, \ldots, \hat{x}_n)$. This means we can test $(\hat{x}_1, \ldots, \hat{x}_n)$ for efficiency in VMP by testing it for optimality in LP* $(\hat{x}_1, \ldots, \hat{x}_n)$. This is most easily performed using the CS conditions.

For example, to test (2, 0, 0) for efficiency in P1, we must examine

$$
\begin{array}{lll}
\text{LP}^*(2,0,0)\text{: maximise} & 2x_1 - 2x_2 + 3x_3 & \\
\text{subject to} & -x_1 + x_2 - x_3 \leqslant -2 & \\
& x_1 + 2x_2 - x_3 \leqslant 2 & \\
& -2x_1 - x_2 - x_3 \leqslant -4 & \\
& x_1 + 3x_2 + x_3 \leqslant 5 & (\ldots + s_1 = 5) \\
& 2x_1 + x_2 + 2x_3 \leqslant 4 & (\ldots + s_2 = 4) \\
& x_1, x_2, x_3 \geqslant 0 &
\end{array}
$$

where we have written all the constraints in '$\leqslant$' form. The dual problem is DP.

$$\begin{array}{lll} \text{DP:} & \text{minimise} & -2v_1 + 2v_2 - 4v_3 + 5y_1 + 4y_2 \\ & \text{subject to} & -v_1 + v_2 - 2v_3 + y_1 + 2y_2 \geqslant 2 \\ & & v_1 + 2v_2 - v_3 + 3y_1 + y_2 \geqslant -2 \\ & & -v_1 - v_2 - v_3 + y_1 + 2y_2 \geqslant 3 \\ & & v_1, v_2, v_3, y_1, y_2 \geqslant 0 \end{array}$$

where $v_k$ is the dual variable associated with the constraint of LP* $(\hat{x}_1, \ldots, \hat{x}_n)$ corresponding to the $k$th objective function and $y_i$ is associated with the $i$th constraint of the original VMP.

To apply the CS conditions we observe that $\hat{x}_1 > 0, \hat{x}_2 = \hat{x}_3 = 0, s_1 = 3 > 0$ and $s_2 = 0$, where $s_1$ and $s_2$ are the slack variables for the original inequalities of P1 which appear in LP* (2, 0, 0) as shown above. The slack variables for the constraints of LP* (2, 0, 0) corresponding to objective functions of P1 are all zero (by definition). If the CS conditions are to be satisfied, we must have $y_1 = t_1 = 0$, which means that (2, 0, 0) is optimal in LP* (2, 0, 0) if there is a non-negative solution of

$$\begin{aligned} -v_1 + v_2 - 2v_3 + 2y_2 \qquad &= 2 \\ v_1 + 2v_2 - v_3 + y_2 - t_2 &= -2 \\ -v_1 - v_2 - v_3 + 2y_2 - t_3 &= 3 \end{aligned}$$

where $t_j$ is the $j$th dual slack variable. Unfortunately, we cannot simply solve these equations and see if the resulting solution is non-negative, as in the examples of Chapter 5, because we have more variables than equations – a consequence of (2, 0, 0) being a degenerate solution of LP* (2, 0, 0) (by construction). We must resort to Phase I of the two-phase method: add an artificial variable and minimise the infeasibility which, initially, is $z^{\mathrm{I}} = -a_1 + t_3$. This yields the solution $(v_1, v_2, v_3) = (0, 0, 1)$, $(y_1, y_2) = (0, 2)$ and shows that (2, 0, 0) *is* optimal in LP* (2, 0, 0) and is therefore efficient in P1.

To test $(0, 1\frac{1}{5}, 1\frac{2}{5})$ for efficiency in P1, we observe that $x_2, x_3 > 0, x_1 = s_1 = s_2 = 0$ so that $t_2 = t_3 = 0$ and the constraints of the dual problem of LP* $(0, 1\frac{1}{5}, 1\frac{2}{5})$, of which we seek a non-negative solution, become

$$\begin{aligned} -v_1 + v_2 - 2v_3 + y_1 + 2y_2 - t_1 &= 2 \\ v_1 + 2v_2 - v_3 + 3y_1 + y_2 \qquad &= -2 \\ -v_1 - v_2 - v_3 + y_1 + 2y_2 \qquad &= 3. \end{aligned}$$

Applying Phase I shows that there is no non-negative feasible solution and, hence, that $(0, 1\frac{1}{5}, 1\frac{2}{5})$ is not efficient. The optimal solution of LP* $(0, 1\frac{1}{5}, 1\frac{2}{5})$ is $(\frac{3}{5}, 0, 1\frac{2}{5})$ and, as we have proved and can easily be verified directly, it dominates $(0, 1\frac{1}{5}, 1\frac{2}{5})$. In the next section, we will show that it is also efficient. efficient.

## 9.3 EFFICIENCY AND WEIGHTING FACTORS

In Section 9.1, we described the weighting-factor approach to conflict resolution. It is natural to ask whether the optimal solution of LP$(w_1, \ldots, w_p)$ is efficient. We can answer in the affirmative provided the weights are all positive. To prove this, let us suppose that $(x_1^*, \ldots, x_n^*)$ is optimal in LP$(w_1, \ldots, w_p)$ and that $w_1, \ldots, w_p > 0$. If this solution were not efficient we would have, for some $(x_1, x_2, \ldots, x_n)$,

$$\sum_j c_{Kj} x_j > \sum_j c_{Kj} x_j^* \qquad \text{for some } K$$

$$\sum_j c_{kj} x_j \geqslant \sum_j c_{kj} x_j^* \qquad \text{for } k \neq K$$

and multiplying the first of these by $w_K$ and the remainder by $w_k$ and summing gives

$$\sum_k \sum_j w_k c_{kj} x_j > \sum_k \sum_j w_k c_{kj} x_j^*$$

which would contradict the assumed optimality of $(x_1^*, \ldots, x_n^*)$. So this solution must be efficient.

For example, in P1, the objective function of LP(2, 1, 3) is $7x_1 - x_2 + 6x_3$ and this gives the optimal solution (2, 0, 0) which must be efficient.

The condition of positive weighting factors is no real restriction, for the assumption that the objectives are desirable means that only positive weights make sense.

A similar argument to that described shows that an optimal solution of LP$^*(\hat{x}_1, \ldots, \hat{x}_n)$ is efficient. For, if $(x_1^*, \ldots, x_n^*)$ were optimal and dominated by the feasible (in VMP) solution $(x_1, \ldots, x_n)$, then

$$\sum_j c_{kj} x_j \geqslant \sum_j c_{kj} x_j^* \geqslant \sum_j c_{kj} \hat{x}_j \qquad \text{for all } k$$

so that $(x_1, \ldots, x_n)$ is also feasible in LP$^*(\hat{x}_1, \ldots, \hat{x}_n)$. It is then easy to see, as above, that the dominance of $(x_1, \ldots, x_n)$ contradicts the optimality of $(x_1^*, \ldots, x_n^*)$.

We have seen that solving LP$(w_1, \ldots, w_p)$ yields efficient solutions and we can ask if all efficient solutions can be obtained in this way. To show that this is true, we will suppose that $(\hat{x}_1, \ldots, \hat{x}_n)$ is feasible and efficient and exhibit $\hat{w}_1, \ldots, \hat{w}_p > 0$ making $(\hat{x}_1, \ldots, \hat{x}_n)$ optimal in LP$(\hat{w}_1, \ldots, \hat{w}_p)$. To achieve this we will compare the dual problem of LP$^*(\hat{x}_1, \ldots, \hat{x}_n)$:

$$\text{D}^*\text{: minimise} \qquad -\sum_k v_k \left(\sum_j c_{kj} \hat{x}_j\right) + \sum_i y_i b_i$$

$$\text{subject to} \quad -\sum_k v_k c_{kj} + \sum_i y_i a_{ij} \geqslant \sum_k c_{kj} \qquad \text{for all } i$$

$$v_k \geqslant 0, \; y_i \geqslant 0 \qquad \text{for all } i \text{ and } k$$

and the dual problem of LP$(w_1, \ldots, w_p)$:

$$\text{D}(w_1, \ldots, w_p)\text{: minimise} \quad \sum_i y_i b_i$$

$$\text{subject to} \quad \sum_i y_i a_{ij} \geqslant \sum_k w_k c_{kj} \qquad \text{for all } j$$

$$y_i \geqslant 0 \qquad \text{for all } i.$$

Since $(\hat{x}_1, \ldots, \hat{x}_n)$ is optimal in LP$^*(\hat{x}_1, \ldots, \hat{x}_n)$, the duality theorem of Chapter 5 states that D$^*$ has an optimal solution, say $(\hat{v}_1, \ldots, \hat{v}_p, \hat{y}_1, \ldots, \hat{y}_m)$ and the optimal objective function values are equal. That is

$$\sum_k \sum_j c_{kj} \hat{x}_j = -\sum_k \hat{v}_k \left( \sum_j c_{kj} \hat{x}_j \right) + \sum_i \hat{y}_i b_i. \tag{9.4}$$

Now the constraints of D$^*$ can be rewritten as

$$\sum_i y_i a_{ij} \geqslant \sum_k (1 + v_k) c_{kj} \qquad \text{for all } j$$

which shows that $(\hat{y}_1, \ldots, \hat{y}_m)$ is feasible in D$(\hat{w}_1, \ldots, \hat{w}_p)$, with $\hat{w}_k = 1 + \hat{v}_k \geqslant 1 > 0$. In addition, (9.4) can be rewritten as

$$\sum_{j,k} (\hat{w}_k c_{kj}) \hat{x}_j = \sum_i \hat{y}_i b_i$$

which says that the objective function values of LP$(\hat{w}_1, \ldots, \hat{w}_p)$ and D$(\hat{w}_1, \ldots, \hat{w}_p)$, its dual problem, are equal. Hence $(\hat{x}_1, \ldots, \hat{x}_n)$ is optimal in LP$(\hat{w}_1, \ldots, \hat{w}_p)$.

This result has many useful consequences, as we shall see in subsequent sections. We shall therefore refer to it as the *Fundamental Theorem of Vector Programming*. To illustrate the theorem, we will see how it enables us to convert any VMP with two objective functions into a parametric programming problem. Consider the problem

$$\begin{array}{lll} \text{P2:} & V \max & -x_1 + 4x_2 \\ & & -2x_1 + x_2 \\ & \text{subject to} & -x_1 + x_2 \leqslant 2 \\ & & -x_1 + 3x_2 \leqslant 9 \\ & & x_2 \leqslant 4 \\ & & x_1, x_2 \geqslant 0. \end{array}$$

The objective function of any LP can be multiplied by any positive constant without affecting the optimal solution(s). This means we can multiply the weighting factors of LP$(w_1, \ldots, w_p)$ by any positive scale factor. Applying this to problems with $p = 2$, we can arrange for $w_1 = 1$ so we have to solve LP$(1, w_2)$. This means we can find all efficient solutions of VMP by solving the parametric programming problem LP$(1, \theta)$ for $\theta > 0$, using the procedures of Chapter 6.

For P2, the objective function of LP$(1, \theta)$ is

$$(-x_1 + 4x_2) + \theta(-2x_1 + x_2) = -(1 + 2\theta)x_1 + (4 + \theta)x_2.$$

When $\theta = 1$, the objective is $-3x_1 + 5x_2$ and maximising this subject to the constraints of P2 leads to the optimal tableau P2/T1 below. However, we have replaced the objective row with the objective rows corresponding to the parametric objective of LP$(1, \theta)$.

| P2 | $s_2$ | $s_1$ | T1 |
|---|---|---|---|
| $x_2$ | $\frac{1}{2}$ | $-\frac{1}{2}$ | $3\frac{1}{2}$ |
| $x_1$ | $\textcircled{\frac{1}{2}}$ | $-\frac{3}{2}$ | $1\frac{1}{2}$ |
| $s_3$ | $-\frac{1}{2}$ | $\textcircled{\frac{1}{2}}$ | $\frac{1}{2}$ |
| $z$ | $1\frac{1}{2}$ | $-\frac{1}{2}$ | $12\frac{1}{2}$ |
| $\theta$ | $-\frac{1}{2}$ | $2\frac{1}{2}$ | $\frac{1}{2}$ |

Tableau P2/T1, corresponding to the BFS $(x_1, x_2) = (1\frac{1}{2}, 3\frac{1}{2})$, is optimal for $\frac{1}{5} \leqslant \theta \leqslant 3$. Pivoting as shown gives BFSs (3, 4), optimal for $-\frac{1}{2} \leqslant \theta \leqslant \frac{1}{5}$, and (0, 2), optimal for $\theta \geqslant 3$. By the fundamental theorem these BFSs constitute the entire list of efficient BFSs. There are also efficient solutions which are not basic. For example, pivoting from $(1\frac{1}{2}, 3\frac{1}{2})$ to (3, 4) corresponds to making $s_1$ basic in P2/T1. Putting $s_1 = \phi \geqslant 0$ in this tableau means that the canonical equations are satisfied by $x_2 = 3\frac{1}{2} + \frac{1}{2}\phi, x_1 = 1\frac{1}{2} + 1\frac{1}{2}\phi$, $s_3 = \frac{1}{2} - \frac{1}{2}\phi$. These solutions are feasible provided $0 \leqslant \phi \leqslant 1$. (Note that $\phi = 1$ corresponds to the BFS (3,4).) When $\theta = \frac{1}{5}$, the objective row coefficient of $s_1$ is zero so that all these solutions have the same (optimal) objective function value. They are all alternative optimal solutions of LP$(1, \frac{1}{5})$ and thus efficient. Similarly, by looking at the $s_2$-column of P2/T1, we see that $x_2 = 3\frac{1}{2} - \frac{1}{2}\phi, x_1 = 1\frac{1}{2} - \frac{1}{2}\phi$, $s_3 = \frac{1}{2} + \frac{1}{2}\phi$ are optimal in LP$(1, 3)$ for $0 \leqslant \phi \leqslant 3$. Thus the efficient solutions of P2 can be expressed as

$$(x_1, x_2) = (1\tfrac{1}{2} - \tfrac{1}{2}\phi, 3\tfrac{1}{2} - \tfrac{1}{2}\phi) \quad \text{for } 0 \leqslant \phi \leqslant 3$$

and

$$(x_1, x_2) = (1\tfrac{1}{2} + 1\tfrac{1}{2}\phi, 3\tfrac{1}{2} + \tfrac{1}{2}\phi) \qquad \text{for } 0 \leqslant \phi \leqslant 1.$$

We initiated the solution of P2 by solving LP$(1, \theta)$, but it is possible that this problem is unbounded whereas LP$(1, \theta)$ has a solution for some other values of $\theta$. In this case the technique of Exercise 6 of Chapter 6 can be used to find an initial value of $\theta$ (see Exercise 4).

## 9.4 FINDING AN EFFICIENT SOLUTION

The use of weighting factors provides a natural method for finding an efficient solution of a VMP. We could set $w_k = 1$ for all $k$ and solve LP$(1, \ldots, 1)$ to obtain an efficient solution. Unfortunately, this approach can fail, as the example P3 illustrates.

$$
\begin{array}{lll}
\text{P3:} & V\max & x_1 - x_2 \\
& & x_1 - 6x_2 \\
& & -x_1 + 3x_2 \\
& \text{subject to} & -x_1 + 3x_2 \leqslant 4 \\
& & -x_1 + 4x_2 \leqslant 5 \\
& & x_1 - 9x_2 \leqslant -4 \\
& & x_1, x_2 \geqslant 0.
\end{array}
$$

The objective function of LP(1, 1, 1) is $x_1 - 4x_2$ which makes LP(1, 1, 1) unbounded. However, this does not mean that P3 has no efficient solutions, as we shall see below.

The difficulty we face is to select $w_k > 0$ making LP$(w_1, \ldots, w_p)$ bounded. Now, boundedness of the primal problem is ensured if the dual problem is feasible (assuming primal feasibility). Consequently, our first step is to find a feasible solution of the dual constraints, for some positive weights, if one exists. In order to handle the requirement $w_k > 0$, we use the freedom to multiply weights by positive scale factors to replace it with $w_k \geqslant 1$, which is equivalent to $w_k = 1 + v_k$, $v_k \geqslant 0$. In the case of P3, this means we must examine the constraints of the dual problem of LP$(1 + v_1, 1 + v_2, 1 + v_3)$ which are

$$-y_1 - y_2 + y_3 \geqslant v_1 + v_2 - v_3 + 1 \tag{9.5a}$$

$$3y_1 + 4y_2 - 9y_3 \geqslant -v_1 - 6v_2 + 3v_3 - 4. \tag{9.5b}$$

Using Phase I of the two-phase method, we obtain $(y_1, y_2, y_3) = (0, 0, 0)$, $(v_1, v_2, v_3) = (0, 0, 1\tfrac{1}{3})$ as a non-negative and feasible solution of (9.5), so LP$(1, 1, 2\tfrac{1}{3})$, with objective function $-\tfrac{1}{3}x_1$, is bounded. Indeed, it has the optimal, and therefore efficient, solution $(x_1, x_2) = (0, 1\tfrac{1}{4})$.

If the third objective of P3 ($-x_1 + 3x_2$) is replaced by $x_1 + 3x_2$ to give problem P4, then the inequalities (9.5) are replaced, after rearrangement, by

$$\begin{aligned} v_1 + v_2 + v_3 + y_1 + y_2 - y_3 &\leqslant -3 \\ -v_1 - 6v_2 + 3v_3 - 3y_1 - 4y_2 + 9y_3 &\leqslant 4. \end{aligned}$$

Using Phase I we can see that there is no feasible, non-negative solution of these inequalities. The fundamental theorem then allows us to deduce that VMP has no efficient solutions. (Otherwise, by the theorem, LP$(w_1, \ldots, w_p)$ would have an optimal solution for some set of positive weights and, therefore, the dual problem is feasible for these weights.)

It is possible that, having determined $w_k > 0$ for which the dual problem of LP$(w_1, \ldots, w_p)$ is feasible, we then discover that LP$(w_1, \ldots, w_p)$ itself is infeasible. This means, of course, that VMP is infeasible and therefore has no efficient solutions. Whatever happens, either we obtain an efficient solution or we can deduce that there are no efficient solutions.

This two-stage procedure is unnecessary if we can be sure, *a priori*, that LP$(1, \ldots, 1)$ is bounded. This will certainly be the case if the constraints of VMP define a bounded feasible region. However, testing for boundedness as a general procedure would not be worth while, in general. Nevertheless, there are some cases where it is easy to recognise that the feasible region is bounded. For example, if the problem includes a constraint of the form

$$\sum_j a_{ij} x_j \leqslant b_i \tag{9.6}$$

where all $a_{ij} > 0$, then the feasible region is bounded, since any feasible $x_j$ must satisfy $x_j \leqslant b_i/a_{ij}$. This applies to problem P1 (either constraint) and so we can be sure that LP$(w_1, w_2, w_3)$ has an optimal solution for any $w_1, w_2, w_3$. More generally, if the sum of some or all of the constraints is of the form (9.6) then the same conclusion can be drawn.

## 9.5 EFFICIENT TABLEAUX

In Section 9.4 we saw how to find an efficient solution. However, the great majority of VMPs will possess more than one efficient solution and a complete resolution of the problem means finding all efficient solutions. This can become very complicated, since efficient solutions which are not BFSs are typically involved.

Here, we will confine ourselves to obtaining all efficient BFSs and, as our first task, we will develop a test to determine whether the BFS corresponding to a given tableau is efficient.

To accommodate multiple objectives, we will extend the simplex tableau by including an additional objective row for each objective function. For example, the initial tableau for problem P1 is P1/T1. Each objective function relates to

| P1 | $x_1$ | $x_2$ | $x_3$ | T1 |
|---|---|---|---|---|
| $s_1$ | 1 | 3 | 1 | 5 |
| $s_2$ | (2) | 1 | 2 | 4 |
| $z_1$ | −1 | 1 | −1 | 0 |
| $z_2$ | 1 | 2 | −1 | 0 |
| $z_3$ | −2 | −1 | −1 | 0 |

one objective row in the tableau and, as usual, rows define equations, so the $z_2$-row represents

$$z_2 + x_1 + 2x_2 - x_3 = 0.$$

Consequently, these additional rows are subject to the usual pivoting rules and the formulae (3.6) for generating the objective row from the objective function coefficients and the remaining rows are still valid. For example, if we pivot as indicated in P1/T1, we obtain tableau P1/T2.

| P1 | $s_2$ | $x_2$ | $x_3$ | T2 |
|---|---|---|---|---|
| $s_1$ | $-\frac{1}{2}$ | $\frac{5}{2}$ | 0 | 3 |
| $x_1$ | $\frac{1}{2}$ | $\frac{1}{2}$ | 1 | 2 |
| $z_1$ | $\frac{1}{2}$ | $1\frac{1}{2}$ | 0 | 2 |
| $z_2$ | $-\frac{1}{2}$ | $1\frac{1}{2}$ | −2 | −2 |
| $z_3$ | 1 | 0 | 1 | 4 |

We would like to be able to recognise whether the BFS corresponding to P1/T2, $(x_1, x_2, x_3) = (2, 0, 0)$, is efficient. We shall see that this can be done by using a suitable generalisation of the single-objective criterion that the objective row coefficients should be non-negative. We need some new notation to describe extra rows and will write $\gamma_{kj}$ for the element in the $j$th column of the $z_k$-row. Thus, the $z_k$-row can be written:

$$z_k + \sum_j \gamma_{kj} x_{N_j} = \gamma_{ko}$$

where $\gamma_{ko}$ is the objective row element in the resource column (= $z_k$ evaluated at the BFS). Formula (3.6a) says that

$$\gamma_{kj} = \sum_i c_{kB_i} \alpha_{ij} - c_{kN_j} \qquad \text{for } j \geqslant 1, \quad k \geqslant 1 \tag{9.7}$$

where $c_{kB_i}$ ($c_{kN_j}$) is the coefficient of $x_{B_i}$ ($x_{N_j}$) in the $k$th objective function.

Our efficiency test will follow from the relationship between VMP and LP$(w_1, \ldots, w_p)$. Since these problems have the same constraints, we could write down a tableau for LP$(w_1, \ldots, w_p)$ having the same basic and non-basic variables (in the same order) as any given tableau of VMP. If $\gamma_{0j}$ is the element in the $j$th column of the objective row of LP$(w_1, \ldots, w_p)$, then applying (3.6a) to this tableau and using (9.7) gives

$$\alpha_{0j} = \sum_i \sum_k (w_k c_{kB_i})\alpha_{ij} - \sum_k w_k c_{kN_j} = \sum_k w_k \gamma_{kj} \quad \text{for } j \geqslant 1.$$

If this tableau of LP$(w_1, \ldots, w_p)$ is optimal $\left(\sum_k w_k \gamma_{kj} \geqslant 0\right)$ for positive weights, then the BFS is optimal and hence efficient by the results of Section 9.3. The requirement $w_k > 0$ can be translated, by scaling the weights, to $w_k \geqslant 1$ or $w_k = 1 + v_k$, $v_k \geqslant 0$. Therefore, the BFS is efficient if we can find a non-negative solution of

$$-\sum_k v_k \gamma_{kj} \leqslant \sum_k \gamma_{kj} \qquad \text{for all } j. \tag{9.8}$$

In the case of tableau T2, these inequalities read:

$$\begin{aligned} -\tfrac{1}{2}v_1 + \tfrac{1}{2}v_2 - v_3 &\leqslant 1 \\ -1\tfrac{1}{2}v_1 - 1\tfrac{1}{2}v_2 \quad &\leqslant 3 \\ 2v_2 - v_3 &\leqslant -1, \qquad v_1, v_2, v_3 \geqslant 0. \end{aligned}$$

This has a solution $(v_1, v_2, v_3) = (0, 0, 1)$, so the BFS is efficient. We established the same result in Section 9.2 by applying the CS conditions to LP* $(2, 0, 0)$ and it is worth noting that the resulting inequalities typically involve more variables than if (9.8) is used, but (9.8) presumes a knowledge of the tableau as opposed to just knowing the BFS.

If a BFS is non-degenerate, it corresponds to only one tableau and the fundamental theorem allows us to deduce that, if the BFS is efficient, then (9.8) is satisfied. For example, P1/T3 can be obtained by pivoting from P1/T1.

| P1 | $x_1$ | $s_1$ | $s_2$ | T3 |
|---|---|---|---|---|
| $x_2$ | 0 | $\frac{2}{5}$ | $-\frac{1}{5}$ | $1\frac{1}{5}$ |
| $x_3$ | 1 | $-\frac{1}{5}$ | $\frac{3}{5}$ | $1\frac{2}{5}$ |
| $z_1$ | 0 | $-\frac{3}{5}$ | $\frac{4}{5}$ | $\frac{1}{5}$ |
| $z_2$ | 2 | $-1$ | 1 | $-1$ |
| $z_3$ | $-1$ | $\frac{1}{5}$ | $\frac{2}{5}$ | $2\frac{3}{5}$ |

The inequalities (9.7) for P1/T3 are

$$
\begin{aligned}
-2v_2 + v_3 &\leqslant 1 \\
\tfrac{3}{5}v_1 + v_2 - \tfrac{1}{5}v_3 &\leqslant -1\tfrac{2}{5} \\
-\tfrac{4}{5}v_1 - v_2 - \tfrac{2}{5}v_3 &\leqslant 2\tfrac{1}{5}
\end{aligned}
$$

and these have no non-negative solution. Since the BFS is non-degenerate this can only mean that it is not efficient.

We will describe a tableau for which (9.8) has a non-negative solution as an **efficient** tableau. We have seen that the efficiency of a non-degenerate BFS is equivalent to the corresponding tableau being efficient. However, the fundamental theorem says that *any* efficient BFS is optimal in $\mathrm{LP}(w_1, \ldots, w_p)$ for some $w_1, \ldots, w_p > 0$ and so it corresponds to some efficient tableau (Section 3.6). Consequently, if we find all efficient tableaux and write down a list of corresponding BFSs (omitting duplicates, if necessary) we shall have found all efficient BFSs, whether or not degeneracy occurs. In the next section we will describe a method for finding all efficient tableaux.

## 9.6 THE MULTIPLE-OBJECTIVE SIMPLEX METHOD

In Section 9.3, we saw how to find all efficient solutions for VMPs with two objectives. In this section, we will describe a method for constructing all efficient tableaux for general VMPs. The method starts from an efficient tableau and constructs others by pivoting. We will illustrate the procedure by solving P4.

$$
\begin{aligned}
\text{P4:}\quad V\max\quad & x_1 - 2x_2 + 4x_3 && (= z_1) \\
& \phantom{x_1 - {}} 2x_2 - 4x_3 && (= z_2) \\
& x_1 + 4x_2 - 4x_3 && (= z_3) \\
\text{subject to}\quad & -x_1 - 4x_2 + x_3 \leqslant -4 \\
& 2x_1 + 3x_2 - x_3 \leqslant 4 \\
& 2x_1 + 2x_2 - x_3 \leqslant 3 \\
& x_1, x_2, x_3 \geqslant 0.
\end{aligned}
$$

To find an initial efficient tableau, we can use the technique of Section 9.4 to find $w_k > 0$ making $\mathrm{LP}(w_1, \ldots, w_p)$ bounded, solve this LP and then, in the optimal simplex tableau, we can replace the objective row with $p$ new objective rows corresponding to the individual objective functions of VMP. For example, in P4, LP(1, 1, 1) is bounded and the non-objective rows of the optimal tableau are given in the non-objective (first three) rows of P4/T1.

The entries in the $z_k$-rows ($\gamma_{kj}$) can be calculated from formula (9.7) which is just a restatement of (3.6). The result is P4/T1 and we know that this tableau is efficient.

| P4 | $x_1$ | $s_2$ | $x_3$ | T1 |
|---|---|---|---|---|
| $s_1$ | $\frac{5}{3}$ | $\frac{4}{3}$ | $-\frac{1}{3}$ | $1\frac{1}{3}$ |
| $x_2$ | $\frac{2}{3}$ | $\frac{1}{3}$ | $-\frac{1}{3}$ | $1\frac{1}{3}$ |
| $s_3$ | ($\frac{2}{3}$) | $-\frac{2}{3}$ | $-\frac{1}{3}$ | $\frac{1}{3}$ |
| $z_1$ | $-2\frac{1}{3}$ | $-\frac{2}{3}$ | $-3\frac{1}{3}$ | $-2\frac{2}{3}$ |
| $z_2$ | $1\frac{1}{3}$ | $\frac{2}{3}$ | $3\frac{1}{3}$ | $2\frac{2}{3}$ |
| $z_3$ | $1\frac{2}{3}$ | $1\frac{1}{3}$ | $2\frac{2}{3}$ | $5\frac{1}{3}$ |

If we make $x_1$ basic in P4/T1, $s_3$ leaves the basis and we can ask if the tableau obtained by making this pivot will be efficient. We can offer a partial answer to this question by observing that, if we can find $\bar{w}_k > 0$ satisfying $\sum_k \bar{w}_k \gamma_{k1} = 0$ and $\sum_k \bar{w}_k \gamma_{kj} \geqslant 0$ for $j = 2$ and 3, then the BFS of P4/T1 is an alternative optimal solution of LP$(\bar{w}_1, \bar{w}_2, \bar{w}_3)$ and another optimal (efficient) solution can be found by pivoting in column 1 ($x_1$-column). This means that the new tableau will be efficient. In terms of (9.8), if $u_j$ denotes the $j$th slack variable in (9.8), we are seeking a non-negative solution to (9.8) with $u_1 = 0$. We have written out inequalities (9.8) for P4/T1 in tableau format as a subsidiary tableau, P4/T1.1 below.

| P4 | $v_1$ | $v_2$ | $v_3$ | T1.1 |
|---|---|---|---|---|
| $u_1$ | ($\frac{7}{3}$) | $-\frac{4}{3}$ | $-\frac{5}{3}$ | $\frac{2}{3}$ |
| $u_2$ | $\frac{2}{3}$ | $-\frac{2}{3}$ | $-\frac{4}{3}$ | $1\frac{1}{3}$ |
| $u_3$ | $\frac{10}{3}$ | $-\frac{10}{3}$ | $-\frac{8}{3}$ | $2\frac{2}{3}$ |
| $z^{\mathrm{I}}$ | $-\frac{7}{3}$ | $\frac{4}{3}$ | $\frac{5}{3}$ | $-\frac{2}{3}$ |

| P4 | $u_1$ | $v_2$ | $v_3$ | T1.2 |
|---|---|---|---|---|
| $v_1$ | $\frac{3}{7}$ | $-\frac{4}{7}$ | $-\frac{5}{7}$ | $\frac{2}{7}$ |
| $u_2$ | $-\frac{2}{7}$ | $-\frac{2}{7}$ | $-\frac{6}{7}$ | $1\frac{1}{7}$ |
| $u_3$ | $-\frac{10}{7}$ | $-\frac{10}{7}$ | $-\frac{2}{7}$ | $1\frac{5}{7}$ |

We can attempt to force $u_1 = 0$ by maximising the infeasibility $z^{\mathrm{I}} = -u_1$. This is achieved in one simplex pivot in P4/T1.2. So we see that the new tableau obtained by making $x_1$ basic will be efficient. The argument obviously generalises to show that, *if there is a non-negative solution of (9.8) with $u_j = 0$, then any new tableau obtained by pivoting in column $j$ is efficient*. Thus, we now seek to force $u_2 = 0$ in P4/T1.1 or, since it also represents the same inequalities, P4/T1.2. But the reader can check that P4/T1.2 is optimal for minimising $u_2$, so no solution of (9.8) exists with $u_2 = 0$. In general, if $u_j$ is basic in a subsidiary tableau and its row has no positive entries apart from a positive entry in the resource column, then there is no feasible solution with $u_j = 0$. Hence, we cannot have $u_3 = 0$ either. Thus, only pivoting in the $x_1$-column is guaranteed to result in an efficient tableau. Making the indicated pivot gives P4/T2.

| P4 | $s_3$ | $s_2$ | $x_3$ | T2 |
|---|---|---|---|---|
| $s_1$ | $-\frac{5}{2}$ | (3) | ($\frac{1}{2}$) | $\frac{1}{2}$ |
| $x_2$ | $-1$ | 1 | 0 | 1 |
| $x_1$ | $\frac{3}{2}$ | $-1$ | $-\frac{1}{2}$ | $\frac{1}{2}$ |
| $z_1$ | $3\frac{1}{2}$ | $-3$ | $-4\frac{1}{2}$ | $-1\frac{1}{2}$ |
| $z_2$ | $-2$ | 2 | 4 | 2 |
| $z_3$ | $-2\frac{1}{2}$ | 3 | $3\frac{1}{2}$ | $4\frac{1}{2}$ |

Making $s_3$ basic in P4/T2 would return us to P4/T1. We therefore try to force $u_2 = 0$ in the subsidiary tableau P4/T2.1. Since P4/T2.1 is not primal feasible we can maximise $z^{\mathrm{I}} = -u_2 + u_1$ using Phase I rules. Making the indicated pivot gives P4/T2.2.

| P4 | $v_1$ | $v_2$ | $v_3$ | T2.1 |
|---|---|---|---|---|
| $u_1$ | $-\frac{7}{2}$ | 2 | $\frac{5}{2}$ | $-1$ |
| $u_2$ | (3) | $-2$ | $-3$ | 2 |
| $u_3$ | $\frac{9}{2}$ | $-4$ | $-\frac{7}{2}$ | 3 |
| $z^{\mathrm{I}}$ | $-6\frac{1}{2}$ | 4 | $5\frac{1}{2}$ | $-3$ |

| P4 | $u_2$ | $v_2$ | $v_3$ | T2.2 |
|---|---|---|---|---|
| $u_1$ | $\frac{7}{6}$ | $-\frac{1}{3}$ | $-1$ | $1\frac{1}{3}$ |
| $v_1$ | $\frac{1}{3}$ | $-\frac{2}{3}$ | $-1$ | $\frac{2}{3}$ |
| $u_3$ | $-\frac{3}{2}$ | $-1$ | 1 | 0 |

In P4/T2.2, we have $u_2$ and $u_3 = 0$ (simultaneously, as it happens) and so pivoting in the second and third columns of P4/T2 gives efficient tableaux. Making the indicated pivots gives P4/T3 and T4.

| P4 | $s_3$ | $s_1$ | $x_3$ | T3 |
|---|---|---|---|---|
| $s_2$ | $-\frac{5}{6}$ | $\frac{1}{3}$ | $\frac{1}{6}$ | $\frac{1}{6}$ |
| $x_2$ | $-\frac{1}{6}$ | $-\frac{1}{3}$ | $-\frac{1}{6}$ | $\frac{5}{6}$ |
| $x_1$ | $\frac{2}{3}$ | $\frac{1}{3}$ | $-\frac{1}{3}$ | $\frac{2}{3}$ |
| $z_1$ | 1 | 1 | $-4$ | $-1$ |
| $z_2$ | $-\frac{1}{3}$ | $-\frac{2}{3}$ | $3\frac{2}{3}$ | $1\frac{2}{3}$ |
| $z_3$ | 0 | $-1$ | 3 | 4 |

| P4 | $s_3$ | $s_2$ | $s_1$ | T4 |
|---|---|---|---|---|
| $x_3$ | $-5$ | 6 | 2 | 1 |
| $x_2$ | $-1$ | 1 | 0 | 1 |
| $x_1$ | $-1$ | 2 | 1 | 1 |
| $z_1$ | $-19$ | 24 | 9 | 3 |
| $z_2$ | 18 | $-22$ | $-8$ | $-2$ |
| $z_3$ | 15 | $-18$ | $-7$ | 1 |

In P4/T3, the reader may verify that there is no feasible solution of (9.8) with $u_1 = 0$. Since making $s_1$ basic returns us to P4/T2 and making $x_3$ basic leads to P4/T4, no new efficient tableaux can be obtained by pivoting from P4/T3. In P4/T4, there is no pivot in the $s_3$-column, whilst making $s_2$ basic leads to P4/T3 and making $s_1$ basic returns us to P4/T2, so no new efficient tableaux can be obtained by pivoting from P4/T4. We will see below that this means we have found *all* efficient tableaux. Thus, the complete list of efficient BFSs is $(x_1, x_2, x_3) = (0, 1\frac{1}{3}, 0), (\frac{1}{2}, 1, 0), (\frac{2}{3}, \frac{5}{6}, 0), (1, 1, 1)$.

At a general stage of the procedure we will have a list of efficient tableaux and 'process' one of them by examining the non-basic variables in turn. We ignore those which cannot be made basic (no positive pivot) or whose introduction into the basis would lead to a tableau already on our list. We then test columns containing the remaining non-basic variables by checking whether (9.8) has a non-negative solution with $u_j = 0$, for column $j$. We then pivot in any column for which a solution is found and add the resulting efficient (unprocessed) tableau to the list. This step is repeated by choosing a new unprocessed tableau and processing it as above. This continues until the list consists entirely of tableaux which have been processed, which must eventually happen since there are only finitely many possible tableaux.

We will now show that *when all the tableaux in our list have been processed, the list contains all efficient tableaux*. To do this, we will start by describing two efficient tableaux reachable from each other by a single pivot as **linked**, if they are both optimal in $\mathrm{LP}(w_1, \ldots, w_p)$ for the same positive weights. This is equivalent to obtaining one tableau from the other by pivoting in the $j$th column while (9.8) has a non-negative solution with $u_j = 0$. Thus, processing a tableau consists of finding new tableaux linked to the one being tested, and any tableau linked to a member of our list of processed tableaux will also be in the final list. To prove that any efficient tableau is in this list, we will show that any pair of efficient tableaux can be connected by a sequence of linked tableaux. Consequently, starting from any tableau on the list we can connect it to any other efficient tableau by a linked sequence, so the other efficient tableau must also be on the list.

Now, suppose that $\bar{\mathrm{T}}$ and $\hat{\mathrm{T}}$ are optimal tableaux and (9.8) is satisfied for $(\bar{w}_1, \ldots, \bar{w}_p)$ and $(\hat{w}_1, \ldots, \hat{w}_p)$ respectively. Consider the parametric problem $\mathrm{LP}(w_1, \ldots, w_p)$ where

$$w_k = \theta \bar{w}_k + (1-\theta)\hat{w}_k \qquad \text{for } k = 1, \ldots, p. \tag{9.9}$$

When $\theta = 0$, $\hat{\mathrm{T}}$ is optimal and, when $\theta = 1$, $\bar{\mathrm{T}}$ is optimal. Using the parametric programming procedure of Section 6.3, we can connect $\hat{\mathrm{T}}$ to $\bar{\mathrm{T}}$ by a sequence of tableaux, each obtained by pivoting from its predecessor. What is more, pairs of successive tableaux are alternative optima of $\mathrm{LP}(w_1, \ldots, w_p)$ for $w_k$ satisfying (9.9). But this means these tableaux are linked because $\bar{w}_k > 0, \hat{w}_k > 0$ implies $w_k > 0$ for $0 \leqslant \theta \leqslant 1$. This justifies the assertion of the preceding paragraph.

## EXERCISES

1. For each of the following problems and trial solutions, test the trial solution for efficiency and, if it is not efficient, find a dominating efficient solution.

(i) $V\max$

$$\begin{aligned} & x_1 + x_2 \\ & 7x_1 + 5x_2 \end{aligned}$$

subject to

$$\begin{aligned} x_1 + 3x_2 &\leqslant 12 \\ 3x_1 + 2x_2 &\leqslant 15 \\ x_1, x_2 &\geqslant 0. \end{aligned}$$

Test $(x_1, x_2) = (5, 0)$.

(ii) $V\max$

$$\begin{aligned} & x_1 + x_2 \\ & 9x_1 + 5x_2 \end{aligned}$$

$$\begin{aligned} \text{subject to} \quad & x_1 + 3x_2 \leqslant 12 \\ & 3x_1 + 2x_2 \leqslant 15 \\ & x_1, x_2 \geqslant 0. \end{aligned}$$

Test $(x_1, x_2) = (5, 0)$.

(iii) $$\begin{aligned} V\max \quad & -x_1 - 2x_2 + 3x_3 \\ & x_1 + 2x_2 + x_3 \\ & -x_1 + 2x_2 - 7x_3 \\ \text{subject to} \quad & x_1 + 3x_2 + 5x_3 \leqslant 6 \\ & x_1 + x_2 + x_3 \leqslant 2 \\ & x_1, x_2, x_3 \geqslant 0. \end{aligned}$$

Test $(x_1, x_2) = (1, 0, 1)$.

2. Find an efficient BFS of each of the following problems or show that there are no efficient solutions

(i) $$\begin{aligned} V\max \quad & x_1 - x_2 \\ & -x_1 + 2x_2 \\ \text{subject to} \quad & 2x_1 - 3x_2 \leqslant 3 \\ & -x_1 + x_2 \leqslant 4 \\ & x_1, x_2 \geqslant 0. \end{aligned}$$

(ii) $$\begin{aligned} V\max \quad & x_1 - x_2 \\ & -x_1 + 2x_2 \\ \text{subject to} \quad & 2x_1 - 3x_2 \leqslant 3 \\ & 2x_1 - x_2 \leqslant 4 \\ & x_1, x_2 \geqslant 0. \end{aligned}$$

(iii) $$\begin{aligned} V\max \quad & x_1 - 4x_2 \\ & -2x_1 + 3x_2 \\ \text{subject to} \quad & 2x_1 - 3x_2 \leqslant 3 \\ & -3x_1 + 4x_2 \leqslant 2 \\ & 2x_1 - x_2 \leqslant -6 \\ & x_1, x_2 \geqslant 0. \end{aligned}$$

(iv) $$\begin{aligned} V\max \quad & 10x_1 - 3x_2 \\ & -4x_1 + x_2 \\ \text{subject to} \quad & 2x_1 - x_2 \leqslant 2 \\ & 3x_1 - 2x_2 \leqslant 2 \\ & x_1, x_2 \geqslant 0. \end{aligned}$$

3. By comparing their dual problems, show that, if $\mathrm{LP}^*(\hat{x}_1, \ldots, \hat{x}_n)$ is feasible and unbounded, then $\mathrm{LP}(w_1, \ldots, w_p)$ is unbounded for any $w_1, \ldots, w_p > 0$. Deduce that VMP has no efficient solutions and hence show that it is possible to either find an efficient solution of a VMP or show that none exists by solving a single LP. Apply this method to Exercise 2.

4. Use a parametric programming procedure (like that of Section 9.3) to find *all* efficient solutions for those problems of Exercise 2 which have efficient

solutions. Hint: Start your parametric programming with the tableau that gave an efficient solution in Exercise 2.

5. A feasible solution $(\hat{x}_1, \ldots, \hat{x}_n)$ of a VMP is said to be **weakly efficient** if there is no feasible solution $(x_1, \ldots, x_n)$, satisfying $\sum_j c_{kj} x_j > \sum_j c_{kj} \hat{x}_j$ for all $k$. Show that, if $(\hat{x}_1, \ldots, \hat{x}_n)$ is optimal in LP$(w_1, \ldots, w_p)$, where $w_1, \ldots, w_p \geqslant 0$, $\sum w_k = 1$, then $(\hat{x}_1, \ldots, \hat{x}_n)$ is weakly efficient. Deduce that any efficient solution is weakly efficient and by considering the following problem show that there can be weakly efficient solutions which are not efficient.

$$\begin{array}{ll} V\max & x_1 + x_2 \\ & 2x_1 + x_2 \\ \text{subject to} & 3x_1 + 4x_2 \leqslant 22 \\ & 4x_1 + 2x_2 \leqslant 21 \\ & x_1, x_2 \geqslant 0. \end{array}$$

6. Show that, if $(\hat{x}_1, \ldots, \hat{x}_n)$ is uniquely optimal in LP$(w_1, \ldots, w_p)$, where $w_1, \ldots, w_p \geqslant 0$, then $(\hat{x}_1, \ldots, \hat{x}_n)$ is efficient in VMP.

7. Find a tableau corresponding to the canonical form of the following inequalities with basis $\{x_1, x_3\}$.

$$\begin{array}{l} 2x_1 + 3x_2 + 10x_3 \leqslant 9 \\ 2x_1 + x_2 + 2x_3 \leqslant 3 \\ x_1, x_2, x_3 \geqslant 0. \end{array}$$

Use your tableau to test the corresponding BFS for efficiency in the VMP with objectives $-x_1 - x_2 + 3x_3$, $x_1 + x_2 + x_3$ and $-x_1 + x_2 + 7x_3$ and constraints as given above.

8. Find *all* efficient BFSs of problem P1 of Section 9.1.

CHAPTER 10

# Multiple-objective Problems – More Methods

## 10.1 GOAL PROGRAMMING – DEVIATION VARIABLES

It can be argued that decision-makers' aims are often not to maximise profits (or sales etc.) but to achieve a satisfactory level (of profits). This involves specifying a **goal** or **target value** for the objective function. The decision-maker tries to maximise the objective up to the goal value but is not interested in values exceeding the goal. For example, in the manufacturing problem described in Section 9.1, if a goal of 25 is specified for revenue, the manufacturer would seek to maximise revenue if it is below 25 but is not concerned with its value once it exceeds 25. More precisely, we can say that the manufacturer wishes to maximise the minimum of revenue and goal:

$$\min\{x_1 + 0.8x_2 + 0.9x_3, 25\}. \tag{10.1}$$

The extremely dichotomous nature of this criterion, with its abrupt switch from maximising the objective to complete disinterest, is open to objection, but such criteria have found favour with some modellers. However, we are still left with the problem of combining these modified criteria into a single objective. In this section, we will use weighting factors, whereas in the next section we shall describe a different, and more radical, approach.

If, in the manufacturing problem of Section 9.1, we specify goals of 25 on revenue (as above), 30 for sales and 12 for exports and apply weighting factors $w_1, w_2, w_3 > 0$ to the resulting criteria, we arrive at

$$\begin{aligned}
\text{P1: maximise}\quad & w_1 \min\{x_1 + 0.8x_2 + 0.9x_3, 25\} + \\
& + w_2 \min\{x_1 + x_2 + x_3, 30\} + \\
& + w_3 \min\{0.4x_1 + 0.6x_2 + 0.2x_3, 12\} \\
\text{subject to}\quad & 7x_1 + 5x_2 + 6x_3 \leqslant 250 \\
& 6x_1 + 9x_2 + 5x_3 \leqslant 210 \\
& x_1, x_2, x_3 \geqslant 0.
\end{aligned}$$

P1 is not a LP, but we shall see that it can be transformed into one by observing that criterion (10.1) can be rewritten as

$$\min\{x_1 + 0.8x_2 + 0.9x_3 - 25, 0\} + 25 \tag{10.2}$$

and, if we define **deviation variables** $d_1$, $e_1 \geqslant 0$ to satisfy

$$d_1 - e_1 = x_1 + 0.8x_2 + 0.9x_3 - 25,$$

then, provided at least one of $d_1$ and $e_1$ is zero, (10.2) is equal to $-e_1 + 25$. This suggests that P1 can be rewritten as P1*.

$$\begin{aligned}
\text{P1}^*\text{: minimise} \quad & w_1e_1 + w_2e_2 + w_3e_3 \; (-25w_1 - 30w_2 - 12w_3) \\
\text{subject to} \quad & x_1 + 0.8x_2 + 0.9x_3 + e_1 - d_1 = 25 \\
& x_1 + x_2 + x_3 + e_2 - d_2 = 30 \\
& 0.4x_1 + 0.6x_2 + 0.2x_3 + e_3 - d_3 = 12 \\
& 7x_1 + 5x_2 + 6x_3 \leqslant 250 \\
& 6x_1 + 9x_2 + 5x_3 \leqslant 210 \\
& x_j, e_k, d_k \geqslant 0 \quad \text{all } j, k.
\end{aligned}$$

Our formulation will be justified provided any optimal solution of P1* has $d_k$ or $e_k = 0$ (or both) for each $k$. But, if we had a feasible solution with $d_k, e_k > 0$, we could reduce $d_k$ and $e_k$ by equal amounts without affecting feasibility and thus reduce the objective function, since $w_k > 0$. Hence, such a solution could not be optimal. Note that this argument relies on positive weighting factors and fails if any $w_k$ is negative.

In some cases, the decision-maker's aim may be to force an objective as close to its goal as possible, both shortfall and excess being undesirable. For example, in our manufacturing example, increasing exports reduces home sales and so we might use the criterion that exports should be as close as possible to their goal: 12. This means that the objective function of P1 is changed to

$$\begin{aligned}
& w_1 \min\{x_1 + 0.8x_2 + 0.9x_3, 25\} + w_2 \min\{x_1 + x_2 + x_3, 30\} \\
& \quad + w_3 \,|0.4x_1 + 0.6x_2 + 0.2x_3 - 12|
\end{aligned}$$

where $|\alpha|$ is the absolute value of $\alpha$ ($=\alpha$ if $\alpha \geqslant 0$ and $= -\alpha$ if $\alpha < 0$). Now, if $d_3$ and $e_3$ satisfy the constraints of P1* and at most one of $d_3$, $e_3$ is positive, then

$$|0.4x_1 + 0.6x_2 + 0.2x_3 - 12| = d_3 + e_3.$$

Consequently, we only have to change the objective function of P1* to

$$w_1e_1 + w_2e_2 + w_3e_3 + w_3d_3$$

(omitting constant terms). By allowing $e_3$ and $d_3$ to have different (positive) coefficients, we could penalise under- and over-achievement differently.

## 10.2 GOAL PROGRAMMING – PRIORITIES

In this section we will assume that $p$ objective functions are arranged in decreasing order of priority. The exact implications of the ordering chosen will become clear below. We will write $\sum_j c_{kj} x_j$ for the $k$th priority objective ($k = 1$ is the highest priority) and $c_{k0}$ for its goal.

To start, we consider the first (highest priority) objective and try to find a feasible solution $(x_1, \ldots, x_n)$ satisfying $\sum_j c_{1j} x_j \geqslant c_{10}$ and, if our search is successful, we impose this inequality as an extra constraint and then turn to the second objective. If no such solution can be found, we impose the constraint that the first objective function should not drop below its optimal value before turning to the second objective. If $(\hat{x}_1, \ldots, \hat{x}_n)$ is optimal for the first objective function, this implies that, if $\sum_j c_{1j} \hat{x}_j < c_{10}$, then we impose the constraint

$$\sum_j c_{1j} x_j \geqslant \sum_j c_{1j} \hat{x}_j.$$

Both cases can be covered by imposing the additional constraint

$$\sum_{j=1} c_{1j} x_j \geqslant \min \left\{ c_{10}, \sum_j c_{1j} \hat{x}_j \right\} \tag{10.3}$$

and then turning to the second objective. A similar procedure is adopted for the second objective, imposing the constraint that it should not drop below whichever is the smaller of its optimal value and its goal. We then proceed to the third objective and so on and continue until all objectives have been considered. The additional constraint imposed on the $k$th objective before proceeding to the $(k + 1)$th is that

$$\sum_j c_{kj} x_j - c_{k0} \geqslant \min \left\{ 0, \sum_j c_{kj} \hat{x}_j - c_{k0} \right\}. \qquad 10.4)$$

Rewriting inequalities such as (10.3) in this form shows that by allowing a constant term in the objective we can effectively make the goal zero for all objectives.

The procedure outlined above can be re-expressed in terms of the criteria, such as (10.1) introduced in Section 10.1. In that section, we combined the criteria additively using weighting factors. Here, we sequentially optimise the criteria, starting with the highest priority objective and imposing (10.4), which says that we do not permit any reduction in the $k$th criterion, when passing from the $k$th to $(k + 1)$st criterion. More generally, we could use weighting factors *and* priorities. However, deviation variables always allow us to reformulate the problem in the form outlined above, which we shall refer to as a **Priority Goal Programming Problem** (PGP).

In the rest of this section we will show, by example, how PGPs may be solved by slightly modifying the simplex method. We will start with problem P2 in which we write '$G$ max' to signal a PGP and list the objectives in order of priority (highest first). Note that the goals are 16, −30 and 30, respectively.

$$\begin{array}{llrl} \text{P2:} & G\max & 4x_1 + 9x_2 + 8x_3 - 16 & (= z_1) \\ & & -29x_1 - 18x_2 - 16x_3 + 30 & (= z_2) \\ & & 7x_1 + 15x_2 + 14x_3 - 30 & (= z_3) \\ & \text{subject to} & 2x_1 + 3x_2 + 2x_3 \leqslant 5 & \\ & & 5x_1 + 2x_2 + 2x_3 \leqslant 4 & \\ & & x_1, x_2, x_3 \geqslant 0. & \end{array}$$

We commence by maximising the first objective function, using the simplex method, until we reach a tableau in which the objective function value is non-negative (so that the goal is achieved) or which is optimal. The initial tableau is P2/T1 and, after two iterations, P2/T2 is reached.

| P2 | $x_1$ | $x_2$ | $x_3$ | T1 |
|---|---|---|---|---|
| $s_1$ | 2 | ③ | 2 | 5 |
| $s_2$ | 5 | 2 | 2 | 4 |
| $z_1$ | −4 | −9 | −8 | −16 |

| P2 | $x_1$ | $s_1$ | $s_2$ | T2 |
|---|---|---|---|---|
| $x_2$ | −3 | 1 | −1 | 1 |
| $x_3$ | $\frac{11}{2}$ | −1 | $\frac{3}{2}$ | 1 |
| $z_1$ | 13 | 1 | 3 | 1 |

In P2/T2 the goal has been achieved, so we impose the additional constraint $z_1 \geqslant 0$. This means we simply consider the $z_1$-row as an extra constraint row and add a row for $z_2$, using the formula (3.6a) for $j \geqslant 0$ (taking $c_{N_0}$ to be $c_0 = 30$). This gives P2/T3. One further pivot results in P2/T4.

| P2 | $x_1$ | $s_1$ | $s_2$ | T3 |
|---|---|---|---|---|
| $x_2$ | −3 | 1 | −1 | 1 |
| $x_3$ | $\frac{11}{2}$ | −1 | $\frac{3}{2}$ | 1 |
| $z_1$ | 13 | 1 | ③ | 1 |
| $z_2$ | −5 | −2 | −6 | −4 |

| P2 | $x_1$ | $s_1$ | $z_1$ | T4 |
|---|---|---|---|---|
| $x_2$ | $\frac{4}{3}$ | $\frac{4}{3}$ | $\frac{1}{3}$ | $1\frac{1}{3}$ |
| $x_3$ | −1 | $-\frac{3}{2}$ | $-\frac{1}{2}$ | $\frac{1}{2}$ |
| $s_2$ | $\frac{13}{3}$ | $\frac{1}{3}$ | $\frac{1}{3}$ | $\frac{1}{3}$ |
| $z_2$ | 21 | 0 | 2 | −2 |

P2/T4 is optimal but $z_2 < 0$ so the goal has not been achieved. The best we can do is $z_2 = -2$. Therefore, we add the constraint $z_2 \geqslant -2$ or, equivalently, $z_2' \geqslant 0$ where $z_2 = z_2' - 2$. Making this substitution in P2/T4 and adding a row for $z_3$ gives P2/T5. Two further pivots give P2/T6.

| P2 | $x_1$ | $s_1$ | $z_1$ | T5 |
|---|---|---|---|---|
| $x_2$ | $\frac{4}{3}$ | $\frac{4}{3}$ | $\frac{1}{3}$ | $1\frac{1}{3}$ |
| $x_3$ | $-1$ | $-\frac{3}{2}$ | $-\frac{1}{2}$ | $\frac{1}{2}$ |
| $s_2$ | $\frac{13}{3}$ | $\frac{1}{3}$ | $\frac{1}{3}$ | $\frac{1}{3}$ |
| $z_2'$ | 21 | 0 | (2) | 0 |
| $z_3$ | $-1$ | $-1$ | $-1$ | 0 |

| P2 | $x_1$ | $s_2$ | $z_2'$ | T6 |
|---|---|---|---|---|
| $x_2$ | $-\frac{11}{2}$ | $-4$ | $\frac{1}{2}$ | 0 |
| $x_3$ | 8 | $\frac{9}{2}$ | $-\frac{1}{2}$ | 2 |
| $s_1$ | $\frac{5}{2}$ | 3 | $-\frac{1}{2}$ | 1 |
| $z_1$ | $\frac{21}{2}$ | 0 | $\frac{1}{2}$ | 0 |
| $z_3$ | $22\frac{1}{2}$ | 3 | $\frac{1}{2}$ | $-2$ |

From P2/T6 we see that the solution is $(x_1, x_2, x_3) = (0, 0, 2)$.

A potential difficulty arises if one of the objectives is unbounded as in problem P3.

$$
\begin{aligned}
\text{P3:}\quad G\max\quad & 2x_1 + x_2 - 6 \\
& -3x_1 - x_2 + 5 \\
& x_1 + 3x_2 - 2 \\
\text{subject to}\quad & x_1 - x_2 \leqslant 1 \\
& 3x_1 - 2x_2 \leqslant 4 \\
& x_1, x_2 \geqslant 0.
\end{aligned}
$$

The initial tableau is P3/T1 and two iterations lead to P3/T2.

| P3 | $x_1$ | $x_2$ | T1 |
|---|---|---|---|
| $s_1$ | (1) | $-1$ | 1 |
| $s_2$ | 3 | $-2$ | 4 |
| $z_1$ | $-2$ | $-1$ | $-6$ |

| P3 | $s_1$ | $s_2$ | T2 |
|---|---|---|---|
| $x_1$ | $-2$ | 1 | 2 |
| $x_2$ | $-3$ | 1 | 1 |
| $z_1$ | (−7) | 3 | $-1$ |

In P3/T2, we see $z_1$ is unbounded. This means that there are feasible solutions with $z_1 \geqslant 0$, but if we were to add $z_1 \geqslant 0$ as a constraint by treating the $z_1$-row as a tableau row and incorporating a $z_2$-row, the resulting tableau would not be primal feasible (and may also be dual infeasible). The difficulty arises because there are no *basic* feasible solutions with $z_1 \geqslant 0$. To avoid this difficulty we can first pivot in the objective row and the unbounded ($s_1$) column. This pivot must result in primal feasibility since, if $i = 0$ denotes the

$z$-row, we have $\alpha_{00} < 0$, as the goal is not yet achieved and $\alpha_{0k} < 0, \alpha_{ik} \leqslant 0$ for $i \geqslant 1$, where $k$ is the pivot column, because of the unboundedness. Hence,

$$\alpha'_{00} = \alpha_{00}/\alpha_{0k} > 0$$

$$\alpha'_{i0} = \alpha_{i0} - \frac{\alpha_{ik}\,\alpha_{00}}{\alpha_{0k}} \geqslant \alpha_{i0} \geqslant 0.$$

Apply this result to P3/T2, by pivoting as indicated; we obtain the non-objective rows of P3/T3. Note that the goal is now achieved, since $z_1 = 0$, and we can complete P3/T3 by adding a $z_2$-row. One further pivot, as indicated, gives P3/T4.

| P3 | $z_1$ | $s_2$ | T3 |
|---|---|---|---|
| $x_1$ | $-\frac{2}{7}$ | $\frac{1}{7}$ (circled) | $2\frac{2}{7}$ |
| $x_2$ | $-\frac{3}{7}$ | $-\frac{2}{7}$ | $1\frac{3}{7}$ |
| $s_1$ | $-\frac{1}{7}$ | $-\frac{3}{7}$ | $\frac{1}{7}$ |
| $z_2$ | $1\frac{2}{7}$ | $-\frac{1}{7}$ | $-3\frac{2}{7}$ |

| P3 | $z_1$ | $x_1$ | T4 |
|---|---|---|---|
| $s_2$ | $-2$ | 7 | 16 |
| $x_2$ | $-1$ | 2 | 6 |
| $s_1$ | $-1$ | 3 | 7 |
| $z_2$ | 1 | 1 | $-1$ |

In P3/T4, the goal is not achieved and so we should add the constraint $z_2 \geqslant -1$. However, the objective row of P3/T4 has only positive entries (apart from the objective function value) and thus a unique optimal solution. This means that there is only one feasible solution satisfying $z_2 \geqslant -1$ and, therefore, adding further objectives cannot lead to a new solution. Consequently, we can assert that the solution of P3 is $(x_1, x_2) = (0, 6)$. This would be true no matter what lower priority objectives were involved. Quite generally, if we reach an optimal tableau, with a unique optimal solution, in which the goal is not achieved, the BFS for that tableau solves the PGP and there is no need to consider lower priority objectives.

## 10.3 MAXIMIN PROGRAMMING

One context in which multiple-objective problems arise is when the objective function can be one of several possibilities, which one being determined by external factors beyond the control of the decision-maker. For example, the profits of the manufacturer introduced in Section 9.1 might be

$$z_1 = x_1 + 0.8x_2 + 0.9x_3$$

or

$$z_2 = 1.3x_1 + 0.7x_2 + 0.5x_3$$

or

$$z_3 = 0.7x_1 + 1.1x_2 + 0.9x_2$$

depending on the state of the market in which he sells his goods. This would arise if the prices of the three products were 1, 0.8, 0.9 in the first case, 1.3, 0.7, 0.5 in the second case and 0.7, 1.1, 0.9 in the third. If sales and profits are ignored, the manufacturer is faced with combining the three objectives into one. If he knew the probabilities of the three states of the market, he could compute and then maximise the expected profit. With no knowledge of these probabilities, he might decide to assume the worst and maximise the lowest objective function, that is maximise $\min\{z_1, z_2, z_3\}$ subject to the constraints.

As it stands, this maximin problem is not an LP but it can be made into one by observing that it is equivalent to maximising $\phi$ subject to $\phi \leqslant \min\{z_1, z_2, z_3\}$ and the constraints. Furthermore, $\phi \leqslant \min\{z_1, z_2, z_3\}$ is equivalent to $\phi \leqslant z_1$, $\phi \leqslant z_2$, $\phi \leqslant z_3$, so the problem becomes P4.

P4: maximise $\phi$

subject to

$$\begin{aligned}
x_1 + 0.8x_2 + 0.9x_3 &\geqslant \phi \\
1.3x_1 + 0.7x_2 + 0.5x_3 &\geqslant \phi \\
0.7x_1 + 1.1x_2 + 0.9x_3 &\geqslant \phi \\
7x_1 + 5x_2 + 6x_3 &\leqslant 250 \\
6x_1 + 9x_2 + 5x_3 &\leqslant 210 \\
x_1, x_2, x_3 &\geqslant 0.
\end{aligned}$$

Discrete maximin problems are also of interest. For example, we could imagine a firm (Firm I) deciding whether to launch a new product and faced with three possible actions. Its profits will depend on the action chosen and the state of the market, which it has classified into four categories. These profits are displayed in Table 10.1 in which rows represent actions and columns represent states of the market. 'Assuming the worst' means measuring each action by the smallest entry in the corresponding row. Then each action has value 1 and all actions are equally good.

| | | State of market $j$ | | | |
|---|---|---|---|---|---|
| | Profits | 1 | 2 | 3 | 4 |
| Action $i$ | 1 | 1 | 2 | 2 | 2 |
| | 2 | 2 | 2 | 2 | 1 |
| | 3 | 3 | 1 | 2 | 4 |

Table 10.1

The pessimistic nature of the maximin approach is open to obvious objections. If carried to extremes, no-one would cross the road because of the outside possibility that they might stumble and be killed by a car. However, in practice when using this approach, highly implausible cases are filtered out early and only outcomes with reasonably high probabilities are considered. It may be felt that, when used with care, the maximin approach adequately reflects a conservative philosophy of decision-making.

Returning to the problem of Table 10.1, we have seen that the best value that can be achieved is 1. However, if Firm I tosses a fair coin and uses action 1 if 'heads' results and action 2 otherwise, then the expected profits for each of the four states of the market are $1\frac{1}{2}$, 2, 2, $1\frac{1}{2}$. The smallest of these is $1\frac{1}{2}$. Since $1\frac{1}{2} > 1$, we see that a pessimistic firm can improve its expected profits by using randomised strategies, where a **randomised strategy** specifies a probability $p_i$ that the $i$th action is chosen. (In this context, we will sometimes refer to the original actions as **pure strategies**.) The expected profit, if the first state of the market prevails, is $p_1 + 2p_2 + 3p_3$ with similar expressions for other states of the market. Hence, the maximin approach means maximising

$$\min\{p_1 + 2p_2 + 3p_3, 2p_1 + 2p_2 + p_3, 2p_1 + 2p_2 + 2p_3, 2p_1 + p_2 + 4p_3\}.$$

Since they are probabilities, $p_1$, $p_2$ and $p_3$ must satisfy

$$p_1 + p_2 + p_3 = 1 \quad \text{and} \quad p_1, p_2, p_3 \geqslant 0.$$

The trick used to obtain P4 allows us to rewrite the problem as problem P5.

$$\begin{aligned} \text{P5:} \quad & \text{maximise} \quad \phi \\ & \text{subject to} \quad p_1 + 2p_2 + 3p_3 \geqslant \phi \\ & \qquad 2p_1 + 2p_2 + p_3 \geqslant \phi \\ & \qquad 2p_1 + 2p_2 + 2p_3 \geqslant \phi \\ & \qquad 2p_1 + p_2 + 4p_3 \geqslant \phi \\ & \qquad p_1 + p_2 + p_3 = 1 \\ & \qquad p_1, p_2, p_3 \geqslant 0. \end{aligned}$$

It is clear that the maximum value of $\phi$ in P5 is positive since, for example, $p_1 = p_2 = p_3 = \frac{1}{3}$ is feasible and gives $\phi > 0$. This means we can add the constraint $\phi > 0$ without affecting the optimal solution. Then, if we write $x_i$ for $p_i/\phi$, the fifth constraint becomes

$$\phi^{-1} = x_1 + x_2 + x_3.$$

Note also that maximising $\phi$ is the same as the minimising $\phi^{-1}$. Thus, dividing the remaining constraints by $\phi$, P5 is transformed into P6

$$\begin{aligned} \text{P6:} \quad & \text{minimise} \quad x_1 + x_2 + x_3 \quad (= -z) \\ & \text{subject to} \quad x_1 + 2x_2 + 3x_3 \geqslant 1 \end{aligned}$$

$$2x_1 + 2x_2 + x_3 \geqslant 1$$
$$2x_1 + 2x_2 + 2x_3 \geqslant 1$$
$$2x_1 + x_2 + 4x_3 \geqslant 1$$
$$x_1, x_2, x_3 \geqslant 0.$$

The constraint $\phi > 0$ is satisfied implicitly, since $x_1 + x_2 + x_3 > 0$ for any feasible solution of P6 and so it has been dropped.

The optimal solution of P6 is $(x_1, x_2, x_3) = (\frac{2}{11}, \frac{3}{11}, \frac{1}{11})$.

Hence $\phi = (\frac{2}{11} + \frac{3}{11} + \frac{1}{11})^{-1} = 1\frac{5}{6}$ and the optimal solution for Firm I is $(p_1, p_2, p_3) = \frac{11}{6}\,(\frac{2}{11}, \frac{3}{11}, \frac{1}{11}) = (\frac{1}{3}, \frac{1}{2}, \frac{1}{6})$.

In general we consider a decision-maker having $m$ actions available. There are $n$ states of nature and he receives a return of $a_{ij}$ if he takes action $i$ and the state is $j$. The pessimistic decision-maker prepared to adopt randomised strategies will have to solve the problem LPI*.

LPI*: maximise $\phi$

subject to $\sum_i a_{ij} p_i \geqslant \phi$ for all $j$

$\sum_i p_i = 1$

$p_i \geqslant 0$ for all $i$

where $p_i$ is the probability of taking action $i$. Provided we can guarantee that the optimal objective function value is positive, LPI* can be converted into LPI.

LPI: minimise $\sum_i x_i$

subject to $\sum_i a_{ij} x_i \geqslant 1$ for all $j$

$x_i \geqslant 0$ for all $i$;

If $(x_1^*, \ldots, x_n^*)$ is optimal in LPI, $\nu_{\mathrm{I}}$, the optimal value for Firm I, is $(x_1^* + \ldots + x_n^*)^{-1}$ and the optimal solution is $(p_1, \ldots, p_n) = \nu_{\mathrm{I}}\,(x_1^*, \ldots, x_n^*)$.

The reduction from LPI* to LPI required the assumption that the optimal objective function value of LPI* be positive. In our example, we were able to deduce this because all the $a_{ij}$s were positive. The presence of $a_{ij} \leqslant 0$ may prevent us from drawing such a conclusion. In such a case, we may use the result that, if $a'_{ij} = a_{ij} + K$ for all $i, j$, then

$$\sum_i a'_{ij} p_i = \sum_i a_{ij} p_i + K \sum_i p_i = \sum_i a_{ij} p_i + K$$

so that comparisons between strategies are unaffected. This means that the optimal action(s) for a pessimistic decision-maker is (are) unaffected and the optimal value is increased by $K$. If we choose $K$ large enough to ensure that $a_{ij} > 0$ for all $i, j$, we can then use LPI to solve the problem. In the example of Table 10.2, we can add 3 to all entries. This gives Table 10.1, so the optimal solution is still $(p_1, p_2, p_3) = (\frac{1}{3}, \frac{1}{2}, \frac{1}{6})$ but the optimal value $\nu_{\text{I}}$ is $1\frac{5}{6} - 3 = -1\frac{1}{6}$.

| | | *j* 1 | 2 | 3 | 4 |
|---|---|---|---|---|---|
| | 1 | −2 | −1 | −1 | −1 |
| *i* | 2 | −1 | −1 | −1 | −2 |
| | 3 | 0 | −2 | −1 | 1 |

Table 10.2

## 10.4 TWO-PERSON ZERO-SUM GAME THEORY

It is possible that the 'state of the market' faced by Firm I in the problem discussed in Section 10.3 is actually determined by a second firm, Firm II (duopoly). Firm I's pessimism may be more readily justified if the competition between the firms is so intense that Firm II's choice of action is solely motivated by the desire to minimise the profits made by Firm I. This supposition is obviously too extreme to be realistic, but its relaxation involves further problems into which we shall not enter here. More generally, we consider two players, I and II, with actions $i(= 1, \ldots, m)$ and $j(= 1, \ldots, n)$, respectively, available. A decision problem in which I receives $a_{ij}$ and II receives $-a_{ij}$, if I plays $i$ and II plays $j$, is called a **two-person zero-sum game** (ZSG – 'zero sum' because the sum of the receipts of the two players is zero).

We have seen how I can solve the problem of choosing a pessimistic strategy, but the analysis assumed that one of the $n$ states of nature will occur or, in game terms, that player II plays one of his pure strategies. However, if I does well to use a randomised strategy, it seems sensible for II to do likewise and we must ask whether LPI* and LPI are still valid if II is using randomised strategies. To analyse this case we note that I assigns a value to the randomised strategy $(p_1, \ldots, p_n)$ of

$$\min_q \sum_{i,j} p_i a_{ij} q_j = \min_q \sum_j b_j q_j$$

where $(q_1, \ldots, q_m)$ is the strategy adopted by II, $\min_q$ indicates the operation of minimising over $q_1, \ldots, q_n \geqslant 0$ satisfying $\sum_j q_j = 1$, and $b_j = \sum_i p_i a_{ij}$. Now, for any randomised strategy $(q_i, \ldots, q_n)$, we have

$$\sum_j b_j q_j \geqslant \sum_j (\min_k b_k) q_j = (\min_k b_k) \sum_j q_j = \min_k b_k.$$

Conversely, by putting $q_k = 1$ and $q_j = 0$ for $j \neq k$, we see that

$$\min_q \sum_j b_j q_j \leqslant b_k \qquad \text{for any } k.$$

So

$$\min_q \sum_j b_j q_j \leqslant \min_k b_k.$$

We have shown that

$$\min_q \sum_{i,j} p_i a_{ij} q_j = \min_j \sum_i p_i a_{ij}$$

which implies that whether player II uses randomised or pure strategies is irrelevant to I's evaluation of his own strategies. So LPI* and LPI are still valid.

There is a certain symmetry about ZSGs. What goes for one player should also go for the other. It therefore appears reasonable to perform a maximin analysis for player II. Note that, since player II receives $-a_{ij}$, if he wishes to maximise his own expected receipts, he will aim to *minimise* the expected receipts of I. For example, in the problem of Section 10.3, we can re-interpret the columns of Table 10.1 as the actions available to Firm II, and Firm II seeks to choose the column with the smallest maximum entry.

In general, player II evaluates his randomised strategy $(q_1, \ldots, q_m)$ pessimistically as

$$\max_p \sum_{i,j} p_i a_{ij} q_j = \max_i \sum_j a_{ij} q_j$$

where $\max_p$ indicates maximisation over $p_1, \ldots, p_m \geqslant 0$ satisfying $\sum_i p_i = 1$. He will try to choose $(q_1, \ldots, q_n)$ in order to minimise this evaluation and a similar analysis to that for player I shows that he can achieve this by solving LPII*.

LPII*: minimise $\psi$

subject to $\sum_j a_{ij} q_j \leqslant \psi$ for all $i$

$$\sum_j q_j = 1$$

$$q_j \geqslant 0 \qquad \text{for all } j,$$

and if the minimal value of $\psi$ is positive, LPII* can be converted into LPII.

$$\text{LPII: maximise} \quad \sum_j y_j$$

$$\text{subject to} \quad \sum_j a_{ij} y_j \leqslant 1 \qquad \text{for all } i$$

$$y_j \geqslant 0 \qquad \text{for all } j$$

where $y_j = q_j/\psi$ for all $j$. Note that, if $a_{ij} > 0$ for all $i, j$ or a constant is added to all payoffs to ensure this result, then $\psi > 0$ for all feasible solutions so LPII* can be transformed into LPII.

For the problem of Section 10.3, Firm II must solve P7.

$$\begin{array}{lll} \text{P7:} & \text{maximise} & y_1 + y_2 + y_3 + y_4 \\ & \text{subject to} & y_1 + 2y_2 + 2y_3 + 2y_4 \leqslant 1 \\ & & 2y_1 + 2y_2 + 2y_3 + y_4 \leqslant 1 \\ & & 3y_1 + y_2 + 2y_3 + 4y_4 \leqslant 1 \\ & & y_1, y_2, y_3, y_4 \geqslant 0. \end{array}$$

Now it is clear that P6 and P7 are dual problems and therefore we can read off the optimal solution of P7 from P6/T1, the optimal tableau of P6. Using the idea of dual tableaux, expounded in Chapter 5, we conclude that the optimal solution of P7 is $(y_1, \ldots, y_4) = (\frac{1}{11}, \frac{4}{11}, 0, \frac{1}{11})$. Hence $(\frac{1}{11} + \frac{4}{11} + 0 + \frac{1}{11})^{-1} = 1\frac{5}{6}$, so that the optimal strategy for II is $\frac{11}{6}(\frac{1}{11}, \frac{4}{11}, 0, \frac{1}{11}) = (\frac{1}{6}, \frac{2}{3}, 0, \frac{1}{6})$.

| P6 | $s_2$ | $s_1$ | $s_4$ | T1 |
|---|---|---|---|---|
| $x_2$ | $-\frac{2}{11}$ | $-\frac{6}{11}$ | $\frac{5}{11}$ | $\frac{3}{11}$ |
| $x_1$ | $-\frac{5}{11}$ | $\frac{7}{11}$ | $-\frac{4}{11}$ | $\frac{2}{11}$ |
| $x_3$ | $\frac{3}{11}$ | $-\frac{2}{11}$ | $-\frac{2}{11}$ | $\frac{1}{11}$ |
| $s_3$ | $-\frac{8}{11}$ | $-\frac{2}{11}$ | $-\frac{2}{11}$ | $\frac{1}{11}$ |
| $z$ | $\frac{4}{11}$ | $\frac{1}{11}$ | $\frac{1}{11}$ | $-\frac{6}{11}$ |

Quite generally, LPI and LPII are dual problems and the duality theorem applied to LPI and LPII says that the optimal objective function values are equal and therefore $\nu_{\text{I}} = \nu_{\text{II}}$, where $\nu_{\text{I}}$ ($\nu_{\text{II}}$) is the optimal value for player I (player II). This can be rewritten as

$$\max_p \min_q \sum_{i,j} p_i a_{ij} q_j \left( = \max_p \min_j \sum_i p_i a_{ij} = \nu_{\text{I}} = \nu_{\text{II}} = \min_q \max_i \sum_j a_{ij} q_i \right) = \min_q \max_p \sum_{i,j} p_i a_{ij} q_j$$

in which form it is often known as the *Fundamental Theorem of Game Theory*. The result applies even if we do not have all $a_{ij} > 0$ since the device of adding a constant $K$ to all terms simply adds $K$ to $\nu_{\text{I}}$ and $\nu_{\text{II}}$. Alternatively, we could have used the fact that LPI* and LPII* are dual problems.

Since our analysis assumes pessimistic behaviour, if player I plays $(p_1^*, \ldots, p_m^*)$, the optimal solution of LPI*, he will receive an expected return of *at least* $\nu_{\text{I}}$. If player II plays $(q_1^*, \ldots, q_n^*)$, the optimal solution of LPII*, his expected loss will be *at most* $\nu_{\text{II}}$ and therefore player I's expected return will be at most $\nu_{\text{II}} = \nu_{\text{I}}$. This says that, if both are playing pessimistically, player I is receiving $\nu_{\text{I}}$ and cannot improve on this, so he has no incentive to change his strategy. A similar conclusion can be drawn for player II. We can describe the situation as **stable** in that neither player would gain from a unilateral change of strategy or, more graphically, that provided randomised strategies are used, pessimism is justified in ZSGs.

If we multiply the $j$th constraint of LPI* by $q_j$ and sum over $j$, and the $j$th constraint of LPII* by $p_i$ and sum over $i$, where $(q_1, \ldots, q_n)$ and $(p_1, \ldots, p_m)$ are randomised strategies, we obtain

$$\sum_{i,j} p_i^* a_{ij} q_i \geqslant \nu_{\text{I}} = \nu_{\text{II}} \geqslant \sum_{i,j} p_i a_{ij} q_j^* .$$

Substituting $(q_1^*, \ldots, q_n^*)$ on the left and $(p_1^*, \ldots, p_m^*)$ on the right gives

$$\nu_{\text{I}} = \nu_{\text{II}} = \sum_{i,j} p_i^* a_{ij} q_j^* ,$$

which again shows that neither player can improve his position by unilaterally adopting a different strategy.

## EXERCISES

1. Solve the following PGPs.

(i) $G$ max
$$\begin{aligned} 14x_1 + 10x_2 &- 35 \\ 10x_1 + 4x_2 &- 20 \\ 8x_1 + x_2 &- 30 \end{aligned}$$

(ii) $G$ max
$$\begin{aligned} 2x_1 + x_2 &- 11 \\ -2x_1 + x_2 &- 10 \\ - x_2 &+ 9 \\ x_1 + x_2 &- 8 \end{aligned}$$

subject to $5x_1 + 2x_2 \leqslant 10$ subject to $x_1 - 2x_2 \leqslant 3$

$3x_1 + 4x_2 \leqslant 12$ $-x_1 + x_2 \leqslant 12$

$x_1, x_2 \geqslant 0.$ $x_1, x_2 \geqslant 0.$

2. Solve the following PGP, firstly for $\theta = 0$ and then for all values of $\theta$.

$$
\begin{array}{ll}
G\max & x_1 + x_2 - 10 - \theta \\
& -2x_2 + 1 \\
& x_1 - 1 \\
\text{subject to} & 3x_1 + 2x_2 \leqslant 12 \\
& x_1 + 3x_2 \leqslant 6 \\
& x_1, x_2 \geqslant 0.
\end{array}
$$

3. Rewrite the following problems as LPs. Do *not* solve the LPs.

(i) maximise $5\min\{x_1 + 7x_2, 11\} + 2\min\{3x_1 + 8x_2, 20\} + |2x_1 + x_2 - 5|$

subject to $2x_1 - x_2 \leqslant 12$

$x_1 + x_2 \leqslant 11$

$x_1, x_2 \geqslant 0.$

(ii) maximise $\min\{x_1 + 3x_2, 3x_1 + 2x_2, 5x_1 + x_2\}$

subject to $7x_1 + 9x_2 \leqslant 25$

$x_1, x_2 \geqslant 0.$

(iii) maximise $2\min\{x_1 + 2x_2, x_1 + x_2, 2x_1 + x_2\} + 5|x_1 + 3x_2 - 7|$

subject to $|x_1 + 9x_2| + 2|4x_1 + 3x_2| \leqslant 12$

$x_1 + x_2 + \min\{3x_1 + 2x_2, 2x_1 + 5x_2\} \geqslant 7$

$x_1, x_2 \geqslant 0.$

4. Solve the following ZSG.

| | 1 | 2 | 3 |
|---|---|---|---|
| 1 | 1 | 0 | −1 |
| 2 | −1 | 0 | 1 |
| 3 | −1 | 1 | 0 |

Hence, solve the following game for any value of $\alpha$

| | 1 | 2 | 3 | 4 |
|---|---|---|---|---|
| 1 | 1 | 0 | −1 | 1 |
| 2 | −1 | 0 | 1 | 0 |
| 3 | −1 | 1 | 0 | −1 |
| 4 | 0 | 1 | −1 | $\alpha$ |

5. A ZSG in which $a_{ij} = -a_{ij}$, for all $i, j$, is called **symmetric.** Why? Show that LPI* and LPII* are identical for a symmetric game and deduce that the optimal strategies are the same for both players and that the value of the game is zero. Show also that, if $\sum_i a_{ij} \hat{p}_i \geqslant 0$ for all $j$, where $p_i \geqslant 0$, then $(\hat{p}_1, \ldots, \hat{p}_n)$ is a maximin strategy.

In a certain game each player displays one or two fingers and simultaneously calls 'one' or 'two' – guessing the number of fingers the other player will display. If one player guesses right and the other wrong, the successful player receives from his opponent a reward equal to the total number of fingers shown by both players. Otherwise the reward is zero. Explain why this is a symmetric game and show that it is optimal to display one finger and call 'two' with probability $\frac{7}{12}$ and display two fingers and call 'one' with probability $\frac{5}{12}$.

6. If $(\frac{1}{3}, \frac{2}{9}, 0, \frac{4}{9})$ were a minimax strategy for II in the following game, what would be its value? Use the CS conditions to verify that this *is* a minimax strategy and find a maximin strategy for I.

| | 1 | 2 | 3 | 4 |
|---|---|---|---|---|
| 1 | −5 | 5 | 3 | 1 |
| 2 | 5 | −6 | 1 | −1 |
| 3 | 3 | 1 | −7 | −3 |
| 4 | 1 | −1 | −3 | −8 |

CHAPTER 11

# Integer Programming

---

## 11.1 INTRODUCTION AND ROUNDING

If a linear programming problem is modified so that the variables are restricted to integer values, the result is called an **integer linear programming problem** (ILP). An example is:

$$
\begin{array}{lll}
\text{P1:} & \text{maximise} & 3x_1 + 4x_2 + 7x_3 \\
 & \text{subject to} & x_1 + 3x_2 + 6x_3 \leqslant 13 \\
 & & 2x_1 + 3x_2 + 4x_3 \leqslant 13 \\
 & & x_1, x_2, x_3 \geqslant 0, \text{ integer.}
\end{array}
$$

For ILPs with two variables a graphical method can be devised (see Exercise 1). This is of little use for problems with more than two variables. Another possible technique is **rounding**, which starts from the optimal solution to the LP obtained by dropping the integer requirement on the variables. We will call this LP the **relaxation** of the ILP. For example, the optimal solution of the relaxation of P1 is $(x_1, x_2, x_3) = (3\frac{1}{4}, 0, 1\frac{5}{8})$. We can then round the non-integer components to integer values. For example, if we decide to round down this gives (3, 0, 1). However, the objective function increases if we increase any variable. So we might try rounding up to (4, 0, 2), but this is not feasible. A more general procedure would be to find all solutions obtained by rounding each non-integer variable in either direction (in the example this gives (3, 0, 1), (3, 0, 2), (4, 0, 1), (4, 0, 2)), select the feasible ones (in the example this means (3, 0, 1) and (4, 0, 1)), and choose whichever of these gives the greatest objective function value, in the example: (4, 0, 1). For large problems, a direct application of this method is impracticable. For example, if the optimal solution of the LP has 40 non-integer components, there are $2^{40}$ or about $10^{12}$ solutions to test and this could well take years, even on a very fast computer. Even after all this effort, the resulting solution may not be optimal. For example, (4, 0, 1) is not optimal in P1, as we shall see. Even worse, there are ILPs in which none of the rounded solutions is feasible (Exercise 1). Nevertheless, provided their limitations are understood, certain heuristic methods, which are essentially sophisticated versions of rounding, can have some value for large-scale problems.

## 11.2 BRANCH-AND-BOUND METHOD

The apparently mild condition of integer-valued variables makes an LP much more difficult to solve. One source of this difficulty is the absence of a duality theory, which means that there is no simple test of whether a feasible solution is optimal. The method to be described, called **Branch-and-Bound** (B & B) has proved as effective as any in the solution of ILPs. We will illustrate it by solving P1.

For reasons which will become clear subsequently we will refer to P1 as problem 0. In the course of solving P1, we will create new ILPs by adding upper- and lower-bound constraints to P1. The $k$th such problem created will be called problem $k$ and the relaxation of problem $k$ will be called LP$k$.

It is an elementary observation that, if the optimal solution of the relaxation of an ILP is integer-valued, it is also the optimal solution to the ILP. So we start by solving LP0, the relaxation of P1, and hope that its solution is integer-valued. The optimal tableau is P1/LP0 (ignore parts of the tableau in brackets and the $\sigma_1$-row for the moment) and we see that our hope is not fulfilled.

| | P1 | $s_2$ | $x_2$ | $s_1$ | LP0 | |
|---|---|---|---|---|---|---|
| | $x_3$ | $-\frac{1}{8}$ | $\frac{3}{8}$ | $\frac{1}{4}$ | $1\frac{5}{8}$ | |
| $(x_1')$ | $x_1$ | $\frac{3}{4}$ | $\frac{3}{4}$ | $\left(-\frac{1}{2}\right)$ | $3\frac{1}{4}$ | $(-\frac{3}{4})$ |
| | $z$ | $1\frac{3}{8}$ | $\frac{7}{8}$ | $\frac{1}{4}$ | $21\frac{1}{8}$ | |
| | $\sigma_1$ | $-\frac{3}{4}$ | $\left(-\frac{3}{4}\right)$ | $\frac{1}{2}$ | $-\frac{1}{4}$ | |

Since $x_1 = 3\frac{1}{4}$ in P1/LP0, the corresponding BFS does not solve problem 0. However, any feasible integer solution must satisfy $x_1 \leqslant 3$ or $x_1 \geqslant 4$ so that if we create two new ILPs (i) by adding $x_1 \leqslant 3$ to problem 0 to get problem 1 and (ii) by adding $x_1 \geqslant 4$ to problem 0 to get problem 2, then the optimal solution to problem 0 must also be optimal in problem 1 or problem 2. Let us solve LP2 in the hope of obtaining an optimal solution which is integer-valued. This involves adding the constraint $x_1 \geqslant 4$ to P1/LP0. (The choice of LP2, rather than LP1, was arbitrary. We shall continue to make arbitrary choices in this selection but in the next section we will discuss how such choices may be resolved more sensibly.) We will use the methods of Chapter 7 throughout this chapter, so we write $x_1 = x_1' + 4, x_1' \geqslant 0$ and the RHS of the $x_1$-row of P1/LP0 changes to $3\frac{1}{4} - 4 = \frac{3}{4}$ as indicated in brackets. Making the indicated pivot in the $x_1$-row to restore primal feasibility gives P1/LP2.

| | P1 | $s_2$ | $x_2$ | $x_1'$ | LP2 | |
|---|---|---|---|---|---|---|
| $(x_3')$ | $x_3$ | $\frac{1}{4}$ | $\frac{3}{4}$ | $\frac{1}{2}$ | $1\frac{1}{4}$ | $(-\frac{3}{4})$ |
| | $s_1$ | $-\frac{3}{2}$ | $-\frac{3}{2}$ | $-2$ | $1\frac{1}{2}$ | |
| | $z$ | $1\frac{3}{4}$ | $1\frac{1}{4}$ | $\frac{1}{2}$ | $20\frac{3}{4}$ | |
| | $\sigma_3$ | $-\frac{1}{4}$ | $-\frac{3}{4}$ | $\left(-\frac{1}{2}\right)$ | $-\frac{1}{4}$ | |

LP2 does not have an integer solution because $x_3 = 1\frac{1}{4}$ so we can create two new ILPs, problems 3 and 4, by adding $x_3 \leqslant 1$, $x_3 \geqslant 2$ to problem 2. The solution process is displayed as an **enumeration tree** in Fig. 11.1, in which node $k$ represents problem $k$. The solution so far consists of nodes 0, 1, 2, 3 and 4. (The reader is advised to temporarily cover up the remaining nodes.)

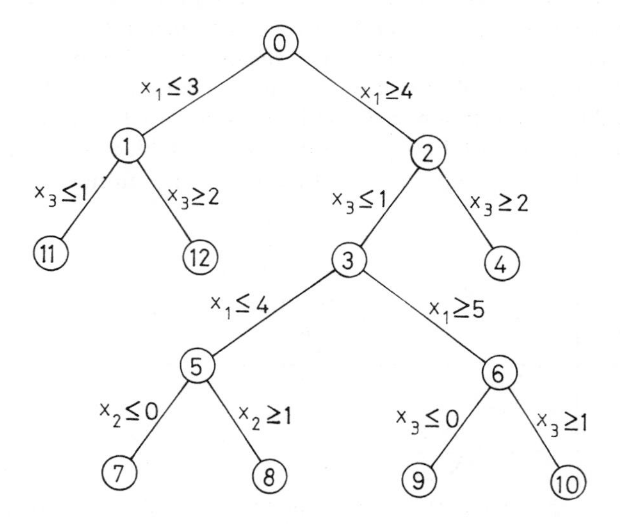


Fig. 11.1 Enumeration tree for P1.

Nodes such as 1, 3 and 4 in the figure, which have not yet been **developed** (by solving the corresponding LP), are called **active.** At each stage we select an active node and develop it. Let us develop node 3, that is, solve LP3. We can do this by making the indicated pivot in the row $\sigma_3$-row of P1/LP2 and the result is P1/LP3.

In P1/LP3, variable $x_1'$ (and therefore $x_1 = x_1' + 4$) is non-integral so we can grow two new branches to nodes 5 and 6 by adding the constraints $x_1' \leqslant 0$ (equivalent to $x_1 \leqslant 4$) or $x_1' \geqslant 1$ (equivalent to $x_1 \geqslant 5$) to problem 3. This gives nodes 0, 1, 2, . . . , 6 of Fig. 11.1, and nodes 1, 4, 5 and 6 are currently active. Developing node 5, by solving LP5, means pivoting as indicated in the $\sigma_1'$-row of

| P1 | $s_2$ | $x_2$ | $\sigma_3$ | LP3 | |
|---|---|---|---|---|---|
| $x_1'$ | $\frac{1}{2}$ | $\frac{3}{2}$ | (−2) | $\frac{1}{2}$ | $(-\frac{1}{2})$ |
| $s_1$ | $-\frac{1}{2}$ | $\frac{3}{2}$ | −4 | $2\frac{1}{2}$ | |
| $z$ | $1\frac{1}{2}$ | $\frac{1}{2}$ | 1 | $20\frac{1}{2}$ | |
| $\sigma_1'$ | $-\frac{1}{2}$ | $(-\frac{3}{2})$ | 2 | $-\frac{1}{2}$ | |

P1/LP3 ($x_1' + \sigma_1' = 0$) and yields the tableau P1/LP5. This time we grow two new branches from node 5 to nodes 7 and 8 to give nodes 0, 1, . . . , 8 of Fig. 11.1. Currently, nodes 1, 4, 6, 7 and 8 are active. We now develop node 7 and find an optimal solution to LP7 by making the indicated pivot in the $\sigma_2$-row in P1/LP5. The reader can verify that this optimal solution is $s_1 = 3$, $s_2 = 1$, $\sigma_2 = \sigma_1' = \sigma_3 = 0$. In terms of the original variables, $(x_1, x_2, x_3) = (4, 0, 1)$. Since this is integral it also solves problem 7. There is therefore no need to grow new branches from node 7. Furthermore, (4, 0, 1) is a feasible solution to P1 and therefore sets a lower bound, $\underline{z} = 19$ on the optimal objective function value of P1. If we discover a better (larger objective function value) solution we will update the value of $\underline{z}$. The best feasible soltuion so far discovered will be called the **incumbent**.

| | P1 | $s_2$ | $\overset{*}{\sigma_1'}$ | $\sigma_3$ | LP5 | |
|---|---|---|---|---|---|---|
| $(x_2')$ | $x_2$ | $\frac{1}{3}$ | $-\frac{2}{3}$ | $(-\frac{4}{3})$ | $\frac{1}{3}$ | $(-\frac{2}{3})$ |
| | $s_1$ | −1 | 1 | −2 | 2 | |
| | $z$ | $1\frac{1}{3}$ | $\frac{1}{3}$ | $1\frac{2}{3}$ | $20\frac{1}{3}$ | |
| | $\sigma_2$ | $(-\frac{1}{3})$ | $\frac{2}{3}$ | $\frac{4}{3}$ | $-\frac{1}{3}$ | |

Before proceeding further we note that in P1/LP5 we have $x_1' \leqslant 0$, i.e. $x_1' + \sigma_1' = 0$. Since all variables are non-negative this imposes the requirement $x_1' = \sigma_1' = 0$. This means that, since $\sigma_1'$ is non-basic in P1/LP5, it can be eliminated and the $\sigma_1'$-column can be dropped, as marked by the asterisk. Consequently, the pivot required to solve LP8 is in the $x_2'$-row and $\sigma_3$-column, as indicated. It is easy to see that, if this pivot is made, the resulting tableau is optimal for LP8 and has objective function value $z_8 = 19\frac{1}{2}$. Since the coefficients in the objective function are integral, the objective function value of any feasible solution of problem 8 is integral and not greater than $19\frac{1}{2}$. Hence this value cannot exceed 19. But $\underline{z} = 19$, so no feasible solution of problem 8 can beat the incumbent. We can therefore declare node 8 **inactive** and no further branching is necessary. In effect, we have pruned out those parts of the tree which could be grown from node 8. This is the 'bound' part of the **B & B** method.

We are now left with active nodes 1, 4 and 6. Developing node 6 means increasing the lower bound on $x_1$ from 4 to 5 in problem 3 and this is achieved by pivoting in the $x_1'$-row in P1/LP3. The result is P1/LP6.

| | P1 | $s_2$ | $x_2$ | $x_1'$ | LP6 | |
|---|---|---|---|---|---|---|
| $(x_3')$ | $x_3$ | $\frac{1}{4}$ | $\frac{3}{4}$ | $\frac{1}{2}$ | $\frac{3}{4}$ | $(-\frac{1}{4})$ |
| | $s_1$ | $-\frac{3}{2}$ | $-\frac{3}{2}$ | $-2$ | $3\frac{1}{2}$ | |
| | $z$ | $1\frac{3}{4}$ | $1\frac{1}{4}$ | $\frac{1}{2}$ | $20\frac{1}{4}$ | |
| | $\sigma_3$ | $-\frac{1}{4}$ | $-\frac{3}{4}$ | $\left(-\frac{1}{2}\right)$ | $-\frac{3}{4}$ | |

P1/LP6 is non-integral, so two new branches are grown with nodes 9 and 10 and the current enumeration tree is shown in Fig. 11.1 if nodes 11 and 12 are ignored.

We now develop currently active nodes 9, 10 and 4. The optimal objective function value of LP9 (pivot in the $\sigma_3$-row of P1/LP6) is $19\frac{1}{2}$ so node 9 can be declared inactive. Adding the constraint $x_3 \geqslant 1$ to problem 6 means that we would like to pivot in the $x_3'$-row of P1/LP6. But there is no pivot available in this row so that LP10, and therefore problem 10, is infeasible. Similarly, the absence of a pivot in the $x_3'$-row of P1/LP2 shows that LP4 is infeasible. The infeasibility of problems 10 and 4 allows us to declare nodes 10 and 4 inactive. This leaves node 1 as the only active vertex. The optimal tableau is P1/LP1, which does not have an integer-valued BFS, so two new branches to nodes 11 and 12 are grown. The current enumeration tree is shown in Fig. 11.1.

| | P1 | $s_2$ | $\sigma_1$ | $s_1$ | LP1 | |
|---|---|---|---|---|---|---|
| $(x_3')$ | $x_3$ | $\left(-\frac{1}{2}\right)$ | $\frac{1}{2}$ | $\frac{1}{2}$ | $1\frac{1}{2}$ | $(-\frac{1}{2})$ |
| | $x_2$ | 1 | $-\frac{4}{3}$ | $-\frac{2}{3}$ | $\frac{1}{3}$ | |
| | $z$ | $\frac{1}{2}$ | $1\frac{1}{6}$ | $\frac{5}{6}$ | $20\frac{5}{6}$ | |
| | $\sigma_3$ | $\frac{1}{2}$ | $-\frac{1}{2}$ | $\left(-\frac{1}{2}\right)$ | $-\frac{1}{2}$ | |

Pivoting in the $\sigma_3$-row of P1/LP1 gives, as optimal solution of LP11, $(x_1, x_2, x_3) = (3, 1, 1)$ with objective function value 20. This exceeds the value of the current incumbent ($\underline{z} = 19$), so $\underline{z}$ is updated to 20 and (3, 1, 1) becomes the new incumbent.

Since the optimal solution of LP12 cannot exceed that of LP1, which is $20\frac{5}{6}$, no solution of problem 12 can beat the incumbent solution and node 12 can be declared inactive. No active nodes remain and we have explored the complete enumeration tree. The optimal solution must be the incumbent solution: (3, 1, 1).

The general procedure is clear. We start with the single node 0, which is considered active, and set $\underline{z}$ large and negative. At any future stage of the method an active node $k$ is selected and developed by solving LP$k$. We will write $z_k$ for the optimal objective function value of LP$k$ and $[z_k]$ for the greatest integer not exceeding $z_k$ (so that $[7\frac{1}{2}] = 7$, $[3] = 3$, $[-5\frac{1}{2}] = -6$).

There are four possibilities for LP$k$:

(i) the optimal solution is not integral and $[z_k] > \underline{z}$;
(ii) the optimal solution is integral and $[z_k] > \underline{z}$;
(iii) $[z_k] \leqslant \underline{z}$;
(iv) LP$k$ is infeasible.

(We have assumed that all objective function coefficients are integers.)

In case (i), two new branches are grown from $k$, giving two new active nodes. In cases (ii), (iii) and (iv), node $k$ is declared inactive. In case (ii) the value of $\underline{z}$ is updated to $[z_k]$ and the incumbent solution replaced by (or defined to be, if there is no current incumbent) the optimal solution of LP$k$. If there are no active nodes left, the problem is solved and the incumbent solution is optimal. Note that in the early stages, when $\underline{z}$ is large and negative and there is no incumbent, the inequality does not need to be checked in cases (i) and (ii).

We have outlined only the bare bones of the B & B method. For example, no mention has been made of which variable to base branching on, or which active node to develop next. How such choices are resolved may have a considerable effect on the efficiency of the method. We will discuss this fleshing-out in more detail in the next two sections.

## 11.3 PENALTIES

In this section we will describe some ways in which the choices involved in applying the B & B method may be resolved. Unfortunately, there is no theory available which points to any particular approach being superior to others. Plausible rules have been developed by studying aspects of ILPs but the ultimate arbiter is practicality: does the rule improve computational efficiency? Such questions can only be answered by experimenting on a computer.

Choices can be made more easily if some of the consequences of such a choice are known. We will therefore examine how bounds can be placed on the change in objective function value when a variable is chosen for branching. We first recall that new nodes are created by choosing a basic structural variable $x_r$ in row $i$ of the optimal tableau of LP$k$ which has a value $\alpha_{io}$ that is not integral. We then create two new problems by adding the constraint $x_r \leqslant [\alpha_{io}]$ or $x_r \geqslant [\alpha_{io}] + 1$ to problem $k$. We will call the first of these the **down-problem** at node $k$ and the second the **up-problem** at node $k$.

The down-problem is solved by adding $x_r \leqslant [\alpha_{io}]$ to the optimal tableau of LP$k$ and using the dual simplex method (in its upper-bound form) to restore optimality. The constraint can be written $x_r + \sigma_r = [\alpha_{io}]$ and involves adding the implicit row (or modifying an existing implicit row to)

$$\sigma_r - \sum_j \alpha_{ij} x_{N_j} = -f_{io}$$

where $f_{io} = \alpha_{io} - [\alpha_{io}] > 0$ is the **fractional part** of $\alpha_{io}$. The first dual-simplex pivot is in column $J$, where $-\alpha_{iJ} < 0$ and

$$\frac{\alpha_{0J}}{\alpha_{iJ}} = \min_{j \geqslant 1} \left\{ \frac{\alpha_{0j}}{\alpha_{ij}} \,\middle|\, (-\alpha_{ij}) < 0 \right\}.$$

The new objective function value, after one pivot in the $\sigma_r$-row and the $J$th column, is

$$\alpha'_{00} = \alpha_{00} - (-f_{io})\alpha_{0J}/(-\alpha_{iJ}).$$

This says that the decrease in objective function value after one dual-simplex pivot is $D_k(x_r)$, where

$$D_k(x_r) = f_{io} \min \left\{ \frac{\alpha_{0j}}{\alpha_{ij}} \,\middle|\, \alpha_{ij} > 0 \right\}$$

and $D_k(x_r)$ is called the **down-penalty**. At every stage of our solution of problem P1 in the preceding section, optimality was restored in a single dual-simplex pivot. When this occurs, the optimal objective function value of the down-problem at node $k$ is $z_k - D_k(x_r)$. In general, several dual-simplex pivots will be required to restore optimality, in which case $z_k - D_k(x_r)$ is an upper bound on the optimal objective function value of the down-problem at node $k$. Note that the formula for the down-penalty (and the up-penalty derived below) would still apply if the variable in row $i$ had been $x'_r$.

We will now re-solve problem P1 of the preceding section. The optimal tableau of the LP relaxation of P1 is, as we saw in Section 11.2, given in P1*/LP0, with the $\sigma_3$-row omitted. (We will refer to P1 as P1* in this section so that tableaux are distinguished from those in Section 11.2.)

The down-penalties are therefore

$$D_0(x_3) = \tfrac{5}{8} \min \left\{ \tfrac{7}{8} / \tfrac{3}{8},\ \tfrac{1}{4} / \tfrac{1}{4} \right\} = \tfrac{5}{8}$$

$$D_0(x_1) = \tfrac{1}{4} \min \left\{ 1\tfrac{3}{8} / \tfrac{3}{4},\ \tfrac{7}{8} / \tfrac{3}{4} \right\} = \tfrac{7}{24}.$$

| P1* | $s_2$ | $x_2$ | $s_1$ | LP0 |
|---|---|---|---|---|
| $x_3$ | $-\frac{1}{8}$ | $\frac{3}{8}$ | $\frac{1}{4}$ | $1\frac{5}{8}$ |
| $x_1$ | $\frac{3}{4}$ | $\frac{3}{4}$ | $-\frac{1}{2}$ | $3\frac{1}{4}$ |
| $z$ | $1\frac{3}{8}$ | $\frac{7}{8}$ | $\frac{1}{4}$ | $21\frac{1}{8}$ |
| $\sigma_3$ | $\frac{1}{8}$ | $-\frac{3}{8}$ | $\left(-\frac{1}{4}\right)$ | $-\frac{5}{8}$ |

The up-problem involves adding $x_r \geqslant [\alpha_{io}] + 1$ to the optimal tableau of LP$k$. This means that $x_r$ is replaced by $x_r' = x_r - [\alpha_{io}] - 1$ and $\alpha_{io}$ is changed to $\alpha_{io} - [\alpha_{io}] - 1 = f_{io} - 1$. The pivot column is $J$, where $\alpha_{iJ} < 0$ and

$$\frac{\alpha_{0J}}{-\alpha_{iJ}} = \min_{j \geqslant 1} \left\{ \frac{\alpha_{0j}}{-\alpha_{ij}} \,\middle|\, \alpha_{ij} < 0 \right\}.$$

This means that the decrease in objective function value after one dual-simplex pivot is $U_k(x_r)$, the **up-penalty**, where

$$U_k(x_r) = (1 - f_{io}) \min_{j \geqslant 1} \left\{ \frac{\alpha_{0j}}{-\alpha_{ij}} \,\middle|\, \alpha_{ij} < 0 \right\}.$$

Once again $z_k - U_k(x_r)$ is an upper bound on the objective function value of the up-problem. In P1*/LP0, we find

$$U_0(x_3) = \tfrac{3}{8} \min \left\{ 1\tfrac{3}{8} / \tfrac{1}{8} \right\} = 4\tfrac{1}{8}$$

$$U_0(x_1) = \tfrac{3}{4} \min \left\{ \tfrac{1}{4} / \tfrac{1}{2} \right\} = \tfrac{3}{8}.$$

We have assumed that a pivot exists for both the up- problem and the down-problem (as was the case in P1*/LP0) so that both up- and down-penalties are defined. We will consider the implications of the failure of this assumption later in the section.

Penalties are used in two ways. Firstly, to choose which variable to branch on. In this context penalties are implicitly treated as if they gave the exact decrease in objective function when a variable is branched on (as opposed to just a bound). Secondly, the bounds provided by penalties can be used to prune the tree and avoid unnecessary pivots.

The use of penalties to choose the bounding variable is motivated by two considerations. Firstly, the desire to find good integer solutions so that the enumeration tree can be severely pruned. Secondly, the hope that branches not developed may be declared inactive by bounding. An empirical rule based on these observations is to branch on the basic structural variable $x_r$ with largest penalty (either up or down). Thus, for P1*/LP0, we would select $x_3$ (since $U_0(x_3)$ is greatest). Two new branches are grown by adding $x_r \leqslant [\alpha_{io}]$ and

$x_r \geqslant [\alpha_0] + 1$ to LP$k$. We then immediately develop the down-problem if $U_k(x_r)$ is the largest penalty and the up-problem if $D_k(x_r)$ is largest. In P1*/LP0 this means growing new branches with $x_3 \leqslant 1$ and $x_3 \geqslant 2$ and developing the down-problem (since $U_0(x_3)$ is greatest). This is achieved by pivoting in the $\sigma_3$-row of P1*/LP0 to get P1*/LP1.

| P1* | $s_2$ | $x_2$ | $\sigma_3$ | LP1 |
|---|---|---|---|---|
| $s_1$ | $-\frac{1}{2}$ | $\frac{3}{2}$ | $-4$ | $2\frac{1}{2}$ |
| $x_1$ | $\frac{1}{2}$ | $\frac{3}{2}$ | $-2$ | $4\frac{1}{2}$ |
| $z$ | $1\frac{1}{2}$ | $\frac{1}{2}$ | $1$ | $20\frac{1}{2}$ |
| $\sigma_1$ | $-\frac{1}{2}$ | ($-\frac{3}{2}$) | $2$ | $-\frac{1}{2}$ |

$$D_1(x_1) = \tfrac{1}{2}\min\left\{1\tfrac{1}{2}/\tfrac{1}{2}, \tfrac{1}{2}/\tfrac{3}{2}\right\} = \tfrac{1}{6}$$

$$U_1(x_1) = \tfrac{1}{2}\min\left\{\tfrac{1}{2}\right\} = \tfrac{1}{4}.$$

Since $U_1(x_1) > D_1(x_1)$, the down-problem at node 1 is developed by adding $x_1 \leqslant 4$, which means pivoting in the $\sigma_1$-row of P1*/LP1 and gives P1*/LP3. The enumeration tree now looks as in Fig. 11.2 if nodes 5 and 6 are ignored.

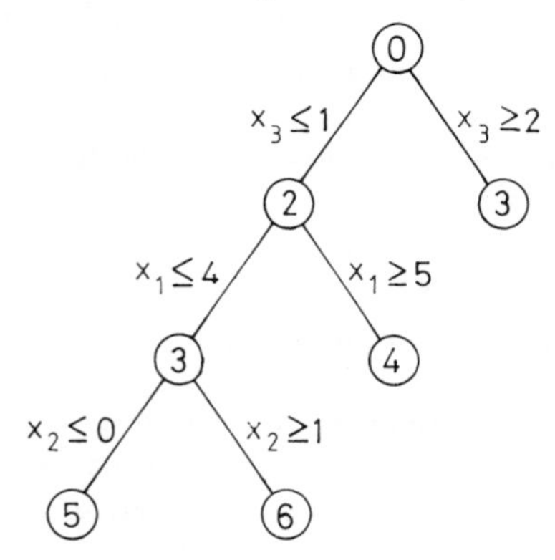


Fig. 11.2 Enumeration tree for P1*.

| | P1* | $s_2$ | $\sigma_1$ | $\sigma_3$ | LP3 | |
|---|---|---|---|---|---|---|
| | $s_1$ | $-1$ | $1$ | $-2$ | $2$ | |
| $(x_2')$ | $x_2$ | $\frac{1}{3}$ | ($-\frac{2}{3}$) | $-\frac{4}{3}$ | $\frac{1}{3}$ | $(-\frac{2}{3})$ |
| | $z$ | $1\frac{1}{3}$ | $\frac{1}{3}$ | $1\frac{2}{3}$ | $20\frac{1}{3}$ | |

$$D_3(x_2) = \tfrac{1}{3}\min\left\{\frac{1\frac{1}{3}}{\frac{1}{3}}\right\} = \frac{4}{3}$$

$$U_3(x_2) = \tfrac{2}{3}\min\left\{\frac{\frac{1}{3}}{\frac{2}{3}}, \frac{1\frac{2}{3}}{\frac{4}{3}}\right\} = \tfrac{1}{3}.$$

The penalties indicate developing the up-problem at node 3 next. We therefore solve LP6 by pivoting in the $x_2'$-row of P1*/LP3 and find an optimal tableau with integer BFS $(x_1, x_2, x_3) = (3, 1, 1)$ and objective function value 20. This becomes the incumbent and we put $\underline{z} = 20$.

Nodes 2, 4 and 5 are currently active and we can now use penalties to prune potential growth of the tree. From the definition of penalties, we have

$$z_5 \leqslant z_3 - D_3(x_2) = 20\tfrac{1}{3} - 1\tfrac{1}{3} = 19 < \underline{z}$$

so node 5 can be declared inactive. Similarly,

$$z_4 \leqslant z_1 - U_1(x_1) = 20\tfrac{1}{2} - \tfrac{1}{4} = 20\tfrac{1}{4}$$

and

$$z_2 \leqslant z_0 - U_0(x_3) = 21\tfrac{1}{8} - 4\tfrac{1}{8} = 17.$$

Hence $[z_4] = 20 \leqslant \underline{z}$, $z_2 < \underline{z}$ so nodes 4 and 2 can be declared inactive and the problem is solved. The complete enumeration tree is shown in Fig. 11.2.

It may happen that the optimal tableau of LP$k$ for some $k$ has $\alpha_{ij} \leqslant 0$ for all $j \geqslant 1$ in a row $i$ containing a structural variable $x_r$ and $\alpha_{i0}$ is not integral. Then $D_k(x_r)$ is not defined and we conventionally put $D_k(x_r) = \infty$. Adding the constraint $x_r \leqslant [\alpha_{i0}]$ involves adding a row with no negative pivot, so the down-problem is infeasible. In this case we can immediately develop the up-problem and declare the node on the down-branch inactive. We will call this a **forced move.** Note that the convention $D_k(x_r) = \infty$ (therefore the largest penalty) makes a forced move consistent with our earlier rule. In the same way, if $\alpha_{ij} \geqslant 0$ for $j \geqslant 1$, we can write $U_k(x_r) = \infty$ and develop the down-problem, declaring the node on the up-branch inactive.

The remaining scope for choice surrounds the selection of an active node for development. The rule we have used in our solutions so far is to develop one of the most recently created active nodes. This rule says that, if a pair of new branches have been created, we immediately develop one of them. Which one can be decided arbitrarily (as in the preceding chapter) or using penalties (as in this chapter). If the nodes are numbered in order of creation, and a node has just been declared inactive (for any reason), the next one to be developed should be the highest numbered active (= most recently created) node. This method is often called **depth-first search** or last-in, first-out (LIFO).

A contrasting rule is called **frontier search.** Whenever the result of the development of node $k$ is that two new branches are grown, the up- and down-nodes are assigned values of $z_k - D_k(x_r)$ and $z_k - U_k(x_r)$ (where $x_r$ is the branching variable) respectively. This implies, *inter alia*, that nodes corresponding to infeasible problems are assigned the value $-\infty$ and can be declared inactive. Then, at each stage, develop the active node with the largest assigned value. This rule tends to reduce the number of nodes in the tree at the expense of storage requirements.

We will illustrate frontier search by solving a **mixed integer linear problem** (MILP) in which some, but not all, of the structural variables are required to be integer-valued. MILPs can be solved using our B & B techniques with two modifications. Firstly, and obviously, we only branch on a variable required to be integer-valued. Secondly, since the optimal objective function value is not necessarily an integer, we can declare a node $k$ inactive by bounding only if $z_k \leqslant \underline{z}$ ($[z_k] \leqslant \underline{z}$ is no longer enough). Practical problems are typically MILPs rather than ILPs.

We will now solve problem P2 which is problem P1 of Section 11.1 modified by requiring only $x_1$ and $x_2$ to be integers. The optimal tableau of the LP relaxation of P2 is P1/LP0 (of Section 11.2). However, $x_3$ is not now required to be an integer so we can only branch on $x_1$. The value assigned to the down-problem (node 1) is $z_0 - D_0(x_1) = 20\frac{5}{6}$ and to the up-problem (node 2) is $z_0 - U_0(x_1) = 20\frac{3}{4}$. Node 1 has the greatest assigned value, so we develop node 1, giving the optimal tableau P2/LP1.

| P2 | $s_2$ | $\sigma_1$ | $s_1$ | LP1 |
|---|---|---|---|---|
| $x_3$ | $-\frac{1}{2}$ | $\frac{1}{2}$ | $\frac{1}{2}$ | $1\frac{1}{2}$ |
| $x_2$ | 1 | $-\frac{4}{3}$ | $-\frac{2}{3}$ | $\frac{1}{3}$ |
| $z$ | $\frac{1}{2}$ | $1\frac{1}{6}$ | $\frac{5}{6}$ | $20\frac{5}{6}$ |

$D_1(x_2) = \frac{1}{6}$, $\text{AV}_3 = 20\frac{2}{3}$

$U_1(x) = \frac{7}{12}$, $\text{AV}_4 = 20\frac{1}{4}$.

In P2/LP1 we branch on $x_2$ to create (down) node 3 with $\text{AV}_3 = 20\frac{2}{3}$ and (up) node 4 with $\text{AV}_4 = 20\frac{1}{4}$, where $\text{AV}_k$ denotes the assigned value at node $k$. The tree now looks like Fig. 11.3 and node 2 has the greatest assigned value of the active nodes (2, 3 and 4) so we solve LP2. This has optimal solution $(x_1, x_2, x_3) = (4, 0, 1\frac{1}{4})$ in which $x_1$ and $x_2$ are integral and $z = 20\frac{3}{4}$. We there-set $\underline{z} = 20\frac{3}{4}$ and $(4, 0, 1\frac{1}{4})$ becomes the incumbent.

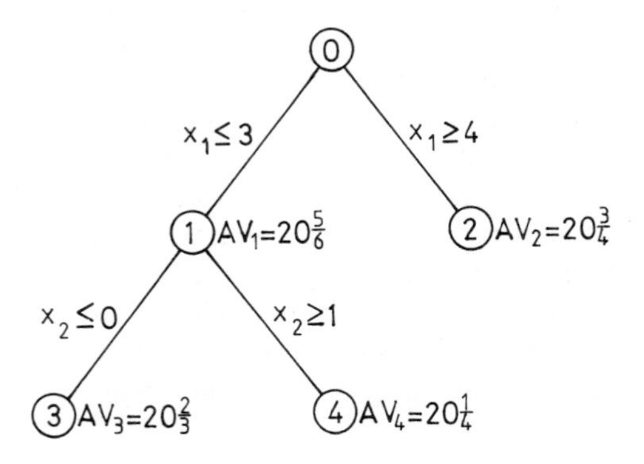


Fig. 11.3 Enumeration tree for P2.

Nodes 3 and 4 remain active. But $z_3 \leqslant AV_3 \leqslant \underline{z}$ and $z_4 \leqslant AV_4 \leqslant \underline{z}$ so 3 and 4 can be declared inactive. The problem is solved and the incumbent solution $(4, 0, 1\frac{1}{4})$ is optimal.

When a choice of branching variable is necessary we can choose the variable which leads to a node having the largest assigned value. This means choosing the variable with smallest penalty (either up or down).

## 11.4 STRATEGIES FOR LARGE-SCALE PROBLEMS

In this section we will offer some suggestions, all of which have proved valuable in practice, intended to render more efficient the implementation of the B & B method on large ILPs and, especially, large MILPs. These suggestions are mainly concerned with ways of making the choices implicit in the B & B methods which were discussed in Section 11.3 and we shall start by considering the selection of a branching variable. No numerical illustrations are given because it is the very nature of the techniques described that they are successful on large problems – too large to illustrate explicitly here.

Experience gained from solving and modelling with ILPs will often suggest that some integer variables in a particular problem are more 'important' than others, in that it is desirable to force the important variables to integer values early in the solution. This is implemented by allowing the solver to specify priorities. When the branching variable is selected, a high priority variable is preferred to one of low priority. Another factor influencing this selection is the observation that variables which are nearly integer-valued often become integers, when branching on other variables. For this reason it can prove profitable to branch whenever possible, only on variables which differ by more than some pre-specified tolerance, 0.1 for example, from the nearest integer.

Even when these two methods are employed, some automatic selection of branching variables is useful. In Section 11.3 we saw how penalties could be used. The justification for selecting the variable with the largest penalty implicitly assumed that the penalty gave a useful estimate of the decrease in objective function value and not just a bound. When only one dual simplex iteration is required to restore optimality, the bound is equal to the decrease, but in large problems, typically, several dual-simplex pivots are needed. If $z_k$ and $z_D$ are the optimal objective function values of, respectively, LP$k$ and the down-problem created by branching on $x_{B_i}$ at node $k$, then we will define the **down-pseudo-cost** of $x_{B_i}$ to be

$$P_D(x_{B_i}) = (z_k - z_D) \,/\, f_{io}$$

where $f_{io}$ is the fractional part of $\alpha_{io}$. The **up-pseudo-cost** is

$$P_U(x_{B_i}) = (z_k - z_U) \,/\, (1 - f_{io})$$

where $z_U$ is the objective function value of the up-problem. Computational experience has shown that pseudo-costs do not vary too much from node to node (which is why we have not given them a '$k$' subscript). Consequently, pseudo-costs can be calculated early in the solution and used to estimate $z_k - z_D$ and $z_k - z_U$ at nodes further down the enumeration tree. When several dual simplex iterations are required to restore optimality, pseudo-costs will generally give better estimates of these quantities than penalties. In addition, pseudo-costs are cheaper to compute. However, unlike penalties, they do not give guaranteed bounds, so penalties and pseudo-costs are not exact competitors.

Forcing $x_r$ to an integer value reduces the objective function by approximately $P_D(x_{B_i})f_{i0}$ for the down-problem and $P_U(x_{B_i})(1 - f_{i0})$ for the up-problem. Hence, branching on $x_{B_i}$ leads to a reduction of at least

$$\min\{P_D(x_{B_i})f_{i0}, P_U(x_{B_i})(1 - f_{i0})\} \tag{11.1}$$

and selecting the variable which maximises (11.1) has proved a successful branching rule.

When solving the up- or down-problem at node $k$ in preceding sections, we first modified the original tableau of LP$k$ to allow for the new or changed, upper or lower bound. Optimality was then restored using the dual simplex method with implicit upper bounds. For large problems, parametric variation of the bound can prove more efficient. Using this technique to change the constraint $x_{B_i} \leqslant U$ to $x_{B_i} \leqslant V(< U)$, for example, would mean solving the parametric problem $x_{B_i} \leqslant U + \theta\,(V - U)$ starting with $\theta = 0$ and terminating at $\theta = 1$. This can be done easily, by combining the methods of Chapters 6 and 7.

Another area of choice in B & B concerns the development rules for exploring the enumeration tree. In Section 11.3, we described two possibilities: depth-first and frontier search. The first has the advantage that no time is wasted in finding a first feasible solution. This means that substantial unprofitable parts of the enumeration tree are often pruned away by bounding. In addition, depth-first search minimises storage requirements, but ignoring objective function values completely does not seem very sensible and this is borne out in practice. Frontier search results in small trees but finds a feasible solution late in the process. This usually means little pruning and, potentially even more serious, since large problems frequently have to be terminated before they have been completely solved to avoid excessive computation time, frontier search may provide no solution at all. A hybrid procedure combining most of the merits of both methods is to develop immediately one of the up- or down-problems (This is depth-first search). However, if at node $k$ no new problems are created (because LP$k$ has an integer solution, or is infeasible, or because of bounding), we develop the active node with the largest 'value'. We shall define this value more precisely below. This rule (and similar variants) produces an optimal solution well before the enumeration tree is fully explored. Since much of the computation time is used to establish that the solution *is* optimal, premature termination of the procedure can often save time and yet still give the optimal solution, although its optimality cannot be guaranteed.

Development rules involving elements of frontier search need to assign a value to active node $k$. Ideally we would set this value equal to the optimal objective function value of problem $k$, say $z_k^*$. The anticipated value used in Section 11.3 and based on penalties is unlikely to be very close to $z_k^*$. We may expect to generate a better estimate by using pseudo-costs. If we make the (unrealistic) assumption that, in driving some variables to integer values, the other variable values do not change much, we can estimate the optimal objective function value of the integer down-problem created by branching on $x_{B_r}$ at node $k$ to be

$$z_k - P_D(x_{B_r}) f_{r0} - \sum_{i \neq r} \min \{P_D(x_{B_i}) f_{i0}, P_U(x_{B_i})(1 - f_{i0})\}$$

where $z_k$ is the optimal objective function value of LP$k$. This expression uses formula (11.1) and a similar expression holds for the up-problem.

The development strategy can be further controlled by specifying a parameter $\gamma$ and temporarily removing nodes satisfying $z_k \leqslant \gamma$ from the list of active nodes. We would then hope that, if $\gamma$ is well chosen, an incumbent will be found that enables node $k$ to be removed permanently. If our choice of $\gamma$ is over-optimistic, we may have to re-instate temporarily inactive nodes. A related approach relaxes the requirement of finding an optimal solution to finding a solution with objective function value within $\epsilon(> 0)$ of the optimum. In the enumeration tree, this means that node $k$ can be declared inactive if $z_k \leqslant \underline{z} + \epsilon$. This often substantially reduces the size of the tree. The resulting solution is called $\epsilon$-optimal.

## 11.5 NON-LINEAR PROGRAMMING USING SPECIAL ORDERED SETS

The B & B approach described in preceding sections can be adapted to a much wider class of problems than integer programming, and in this and the next section we will offer two examples. Both topics covered have been studied in much greater depth than we have space to discuss here, but our intention is only to illustrate the breadth of applicability of the B & B philosophy.

Problems in which the objective function or the constraints (or both) are non-linear are called **Non-linear Programming Problems** (NLPs). We will describe a procedure for solving certain NLPs. Initially, we will illustrate the procedure by 'solving' a very simple, though non-trivial, example: P3 below.

$$\begin{aligned} \text{P3:} \quad & \text{maximise} && f(x_1) - x_2 \\ & \text{subject to} && 4x_1 + x_2 \leqslant 11 \\ & && x_1, x_2 \geqslant 0 \end{aligned}$$

where

$$f(x_1) = x_1^3 - 4x_1^2 + 4x_1 + 2.$$

Rather than solve P3 itself we will solve an approximate version of the problem, generated by approximating $f$. To do this, we evaluate $f$ at certain values of $x_1$ and draw straight lines between adjacent points. This is illustrated in Fig. 11.4 where $f$ has been evaluated at $x_1 = 0, 1, 2, 3$. (There is no need to choose larger or smaller values as they would not be feasible in P3.)

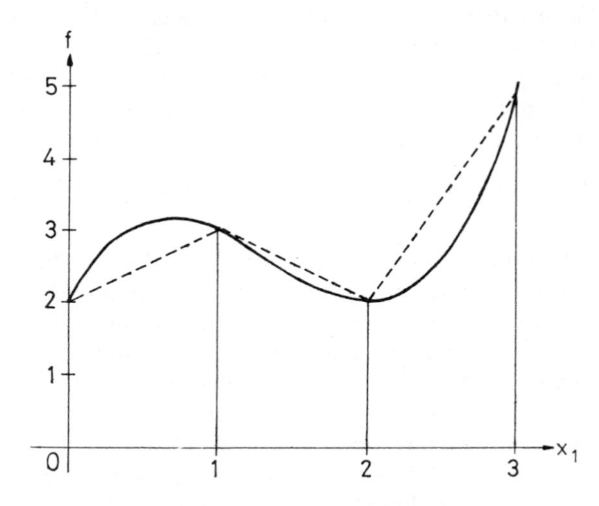


Fig. 11.4 PWL approximation.

The approximating function $\phi$, drawn dashed in the figure, is called a **Piecewise-Linear (PWL) approximation** and we will now define $\phi$ rigorously. For $0 \leqslant x_1 \leqslant 1$, let us write $x_1 = \lambda_2$, where $\lambda_1 + \lambda_2 = 1$ (so that $\lambda_1, \lambda_2 \geqslant 0$), and define

$$\phi(x_1) = \lambda_1 f(0) + \lambda_2 f(1) = 2\lambda_1 + 3\lambda_2$$

since $f(0) = 2, f(1) = 3$. Similarly, for $1 \leqslant x_1 \leqslant 2$, we can write $x_1 = \lambda_2 + 2\lambda_3$, where $\lambda_2 + \lambda_3 = 1$, and define

$$\phi(x_1) = \lambda_2 f(1) + \lambda_3 f(2) = 3\lambda_2 + 2\lambda_3$$

since $f(2) = 2$. The definition for $2 \leqslant x_1 \leqslant 3$ should now be obvious. As it stands, this definition of $\phi$ is rather cumbersome, but it can be restated in a more succint form as follows (noting that $f(3) = 5$),

$$\phi(x_1) = 2\lambda_1 + 3\lambda_2 + 2\lambda_3 + 5\lambda_4,$$

where

$$x_1 = \lambda_2 + 2\lambda_3 + 3\lambda_4, \tag{11.2}$$

and

$$1 = \lambda_1 + \lambda_2 + \lambda_3 + \lambda_4,$$

and $\lambda_1, \ldots, \lambda_4 \geqslant 0$, provided that $\lambda_1, \ldots, \lambda_4$ constitute a **Special Ordered Set** (SOS). This stipulates that either $\lambda_1$, or $\lambda_2$, or $\lambda_3$, or $\lambda_4$, or $\lambda_1$ and $\lambda_2$, or $\lambda_2$ and $\lambda_3$, or $\lambda_3$, and $\lambda_4$ is (are) positive.

In general, $\lambda_1, \ldots, \lambda_p$ constitute an SOS if $\lambda_j = 0$ for all $j \neq k$ or $\lambda_j = 0$ for all $j$ except $k$ and $k + 1$, for some $k$. The general PWL approximation of a function $f(x)$ is specified in terms of $p$ values of $x$: $a_1, \ldots, a_p$, with $a_1 < a_2 < \ldots < a_p$, by

$$\phi(x) = \sum_k \lambda_k f(a_k)$$

where

$$x = \sum_k a_k \lambda_k,$$

and

$$1 = \sum_k \lambda_k,$$

where $\lambda_1, \ldots, \lambda_p \geqslant 0$ constitute an SOS.

We can rewrite P3 using our PWL approximation as

$$\begin{array}{lll}
\text{P3}^*: \text{ maximise} & 2\lambda_1 + 3\lambda_2 + 2\lambda_3 + 5\lambda_4 - x_2 & (= z) \\
\text{subject to} & 4\lambda_2 + 8\lambda_3 + 12\lambda_4 + x_2 \leqslant 11 & \\
 & \lambda_1 + \lambda_2 + \lambda_3 + \lambda_4 = 1 & \\
 & \lambda_1, \ldots, \lambda_4, x_2 \geqslant 0 &
\end{array}$$

where $\lambda_1, \ldots, \lambda_4$ constitute an SOS. This last requirement means that P3* is not an LP but we can handle it using a B & B method, as we now illustrate.

We first solve LP0 (P3* with the SOS requirement relaxed). The optimal solution of this LP is $\lambda_1 = \frac{1}{12}$, $\lambda_4 = \frac{11}{12}$, $\lambda_2 = \lambda_3 = x_2 = 0$ and $z_0 = 4\frac{3}{4}$. Only $\lambda_1$ and $\lambda_4$ are non-zero, but they are not adjacent and so do not satisfy the SOS requirements. Now any SOS solution must have either $\lambda_1 > 0$ (and thus $\lambda_3 = \lambda_4 = 0$) or $\lambda_1 = 0$. Thus we create problem 1 by adding $\lambda_1 = 0$ to P3* and problem 2 by adding $\lambda_3 = \lambda_4 = 0$. For a general SOS, we choose some $k$ and create branches by adding either $\lambda_1 = \ldots = \lambda_{k-1} = 0$ or $\lambda_{k+1} = \ldots = \lambda_p = 0$. Our choice corresponds to $k = 2$. The resulting sets of $\lambda_j$s are still SOSs.

The optimal solution of LP1 is $\lambda_2 = \frac{1}{8}$, $\lambda_4 = \frac{7}{8}$, $\lambda_1 = \lambda_3 = x_2 = 0$ and $z_1 = 4\frac{3}{4}$, so we branch again by adding $\lambda_2 = 0$ (problem 3) and $\lambda_4 = 0$ (problem 4). The optimal solution of LP3 is $\lambda_3 = \frac{1}{4}$, $\lambda_4 = \frac{3}{4}$, $\lambda_1 = \lambda_2 = x_2 = 0$. This satisfies the SOS requirements and so we put $\underline{z} = z_3 = 4\frac{1}{4}$. The optimal solutions of LP4 and LP2 satisfy $z_4 = z_2 = 3 < \underline{z}$, so the enumeration is completed. The optimal solution is that of LP3, which has $x_1 = 2 \times \frac{1}{4} + 3 \times \frac{3}{4} = 2\frac{3}{4}$, by (11.2) and $x_2 = 0$.

The B & B scheme outlined for problems involving SOSs differs from that for ILPs in one notable respect, namely it permits the possibility of the same solution being feasible for both branches created at a node. In particular, if we have $\lambda_1 = \ldots = \lambda_{k-1} = 0$ on one branch and $\lambda_{k+1} = \ldots = \lambda_p = 0$ on the other, then $\lambda_k = 1, \lambda_i = 0$ for $i \neq k$ is an SOS solution satisfying the constraints on both branches. It is responsible, in the solution of P3$^*$, for LP2 and LP4 having identical optimal solutions ($\lambda_2 = 1, \lambda_1 = \lambda_3 = \lambda_4 = x_2 = 0$). Such an occurrence would not be possible when solving ILPs. Although the existence of common solutions to distinct branches does not affect the validity of the B & B procedure, it could conceivably lead to increased computation time. However, the remedies for avoiding common solutions, for example a positivity constraint on one branch, are likely to involve more additional computation time than their use would save. It is simpler to learn to live with the proposed branching rule.

The choices noted in Sections 11.3 and 11.4, which must be resolved when implementing the B & B scheme for ILPs are also relevant when using SOSs. Similar concepts can be employed in their resolution. For example, pseudo-costs can be generalised. In addition, the accuracy of the approximation must be considered, possibly within the B & B solution. In our problem, we would look at how well P3$^*$ approximates P3. For example, we can observe that, at the optimal solution $(x_1, x_2) = (2\frac{3}{4}, 0)$ of P3$^*$, $z$ (in P3$^*$) $= 4\frac{1}{4}$. whilst $f(2\frac{3}{4}) =$ 3.55. But this only gives information at one point. Ideally we would like to know how good the approximation is over all $x$. In addition, we could change the number of points in each SOS. One obvious, and successful, method is to start with a course representation and subsequently make it finer around the optimal solution, using the accuracy of the approximation as a guide. When pursued in detail, such procedures can become quite complicated.

Finally, we must consider what class of NLPs can be treated by PWL approximation using SOSs. In the case of problems with two variables we can approximate the NLP:

$$\begin{aligned} &\text{maximise} && f(x_1) + g(x_2) \\ &\text{subject to} && h_i(x_1) + k_i(x_2) \leqslant b_i \qquad \text{for all } i \\ &&& x_1, x_2 \geqslant 0 \end{aligned}$$

by taking $a_1, \ldots, a_p \geqslant 0, d_1, \ldots, d_q \geqslant 0$ and writing

$$\begin{aligned} &\text{maximise} && \sum_k \lambda_k f(a_k) + \sum_l \mu_l g(d_l) \\ &\text{subject to} && \sum_k \lambda_k h_i(a_k) + \sum_l \mu_l k_i(d_l) \leqslant b_i \qquad \text{for all } i \end{aligned}$$

$$\sum_k \lambda_k = \sum_l \mu_l = 1$$

$$\lambda_k, \mu_l \geqslant 0 \qquad \text{for all } k,$$

where $x_1 = \sum_k a_k \lambda_k$ and $x_2 = \sum_l d_l \mu_l$.

This approximation obviously extends to problems with more than two variables provided the objective function and constraints are **separable**, that is, they are sums of functions of individual variables, for example, $z = \sum_j f_j(x_j)$. Even separability can sometimes be forced on problems. For example, a non-separable term such as $x_1 x_2$ can be written $x_1 x_2 = y_1^2 - y_2^2$, where $y_1 = \frac{1}{2}x_1 + \frac{1}{2}x_2$, $y_2 = \frac{1}{2}x_1 - \frac{1}{2}x_2$ are added as constraints, although the use of such tricks can adversely affect computation time. However, the really important area of application is to problems which would be linear but for a few (separable) non-linear terms and here it can be very effective.

## 11.6 THE TRAVELLING SALESMAN'S PROBLEM

Consider a salesman who has to visit each of $n$ towns once, starting from and returning to town 1. We will call any such trip a **tour**. The Travelling Salesman's Problem (TSP) is to find a minimal length tour, where $c_{ij}$ is the distance from town $i$ to town $j$. Apart from its direct interpretation, the TSP can also be applied to problems that have nothing to do with towns and salesmen. For example, the problem of sequencing the production of $n$ batches of paint of different colours on one machine, where $c_{ij}$ is the cost of cleaning the machine when colour $j$ is to follow colour $i$, can be formulated as a TSP. A great deal of research has been undertaken into the understanding of TSPs and our intention here is simply to illustrate an alternative branching scheme using the special structure of the problem. We will do this by solving an example, leaving the (straightforward) generalisation of the method to the reader.

For any tour, we will write $x_{ij} = 1$ if town $j$ succeeds town $i$ in the tour and $x_{ij} = 0$ otherwise. Then the length of the tour is $\sum_{i,j} c_{ij} x_{ij}$ and this is to be minimised subject to $\{x_{ij}\}$ representing a tour. Any tour must leave town $i$, so $\sum_j x_{ij} = 1$ for each $i$, and arrive at town $j$, so $\sum_i x_{ij} = 1$ for each $j$. These are the objective function and constraints of an AP, but TSP is not an AP. To see this, consider the problem of Table 11.1. Using the method of Section 8.4, we can find the optimal solution of AP0, the corresponding AP. It is $x_{13} = x_{26} = x_{35} = x_{42} = x_{51} = x_{64} = 1$ and this is represented graphically in Fig. 11.5. From this figure we see that the optimal solution of AP is not a tour so that AP is not TSP. However, AP is a relaxation of TSP, but extra constraints eliminating subtours

| $c_{ij}$ | 1 | 2 | 3 | 4 | 5 | 6 |
|---|---|---|---|---|---|---|
| 1 | – | 7 | 3 | 12 | 5 | 8 |
| 2 | 4 | – | 2 | 10 | 9 | 3 |
| 3 | 6 | 7 | – | 11 | 1 | 7 |
| 4 | 7 | 3 | 1 | – | 8 | 8 |
| 5 | 2 | 10 | 2 | 7 | – | 3 |
| 6 | 4 | 11 | 7 | 6 | 3 | – |

Table 11.1

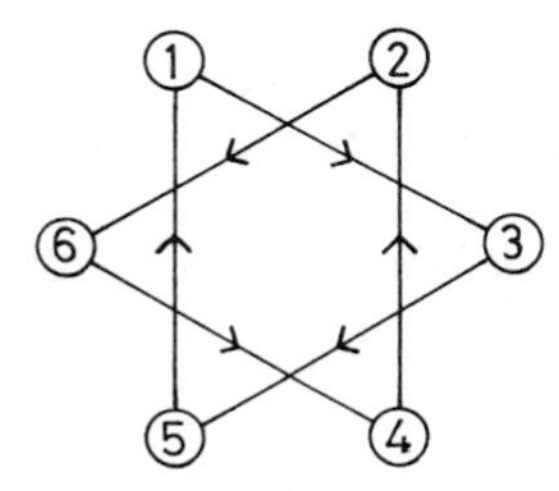


Fig. 11.5 Subtours in the relaxation of a TSP

(for example, $1 \to 3 \to 5 \to 1$ in Fig. 11.5) are required in TSP. We will not write these down explicitly but will impose them successively in a B & B procedure.

We can eliminate the subtour involving towns 1, 3 and 5 by specifying that the tour must go from town 1 to 2, 4 or 6 (problem 1) or from town 3 to 2, 4 or 6 (problem 2) or from town 5 to 2, 4 or 6 (problem 3). In problem 1, we want $x_{12}$ or $x_{14}$ or $x_{16} = 1$ and this can be ensured by imposing $x_{13} = x_{15} = 0$. In problem 2, we want $x_{32}$ or $x_{34}$ or $x_{36} = 1$, and in addition we don't need $x_{12}$ or $x_{14}$ or $x_{16} = 1$, since this is included in problem 1. So problem 2 has $x_{31} = x_{35} = 0$ and $x_{12} = x_{14} = x_{16} = 0$. Similarly, problem 3 has $x_{51} = x_{53} = 0, x_{12} = x_{14} = x_{16} = 0$ and $x_{32} = x_{34} = x_{36} = 0$.

To solve AP$k$, the relaxation of problem $k$, we impose constraints such as $x_{ij} = 0$ by changing $c_{ij}$ to $K$, where $K$ is some large number. Thus, to solve AP1, we change $c_{13}$ and $c_{15}$ to $K$ and start from the optimal solution of AP0. The optimal solution of AP1 is $x_{12} = x_{26} = x_{35} = x_{43} = x_{51} = x_{64} = 1$ and the reader can verify that this corresponds to the tour $1 \to 2 \to 6 \to 4 \to 3 \to 5 \to 1$ with length 20. The optimal solution of AP2 has length 24 and is eliminated by bounding. (This is a *minimisation* problem.) Similarly, the optimal solution of AP3 has length 20 and is also eliminated. Thus, the enumeration tree is completed and the optimal solution of the TSP is the tour of length 20 given above.

## EXERCISES

1. Devise a graphical method for solving ILPs with two variables and illustrate it by solving the following example.

$$\begin{array}{ll} \text{maximise} & 3x_1 + x_2 \\ \text{subject to} & 2x_1 + 3x_2 \leqslant 6 \\ & 2x_1 - 3x_2 \leqslant 3 \\ & x_1, x_2 \geqslant 0, \text{ integer.} \end{array}$$

What happens if rounding is applied to this problem?

2. Solve the following problems in integer variables. Use penalties and depth-first search.

(i)
$$\begin{array}{ll} \text{maximise} & x_1 + 3x_2 \\ \text{subject to} & 2x_1 - 3x_2 \leqslant 4 \\ & -x_1 + 2x_2 \leqslant 7 \\ & 3x_1 + x_2 \leqslant 9 \\ & x_1, x_2 \geqslant 0. \end{array}$$

(ii)
$$\begin{array}{ll} \text{minimise} & -x_1 + 3x_2 - x_3 \\ \text{subject to} & 2x_1 + x_2 - x_3 \leqslant 4 \\ & 4x_1 - 3x_2 \leqslant 2 \\ & -3x_1 + 2x_2 + x_3 \geqslant 3 \\ & x_1, x_2, x_3 \geqslant 0. \end{array}$$

3. Do either of the problems of Exercise 2 have alternative optimal solutions?

4. Solve the following MILP, using frontier search.

$$\begin{array}{ll} \text{maximise} & 3x_1 - 3x_2 + x_3 \\ \text{subject to} & x_1 + x_2 + 2x_3 \leqslant 12 \\ & 2x_1 - 3x_2 - x_3 \leqslant 6 \\ & x_1, x_2, x_3 \leqslant 0, \quad x_1, x_2 \text{ integer.} \end{array}$$

5. Show that, if the LP relaxation of an ILP with integer data is unbounded, so is the ILP.

6. Use a modified B & B method to find an $\epsilon$-optimal solution of the following problem when $\epsilon = 1$. Use penalties and depth-first search.

$$\begin{array}{ll} \text{maximise} & 6x_1 + 3x_2 + 9x_3 + 5x_4 \ (= z) \\ \text{subject to} & x_1 + x_2 + x_3 + x_4 \leqslant 4 \\ & 4x_1 + 2x_2 + 5x_3 + 3x_4 \leqslant 17 \\ & x_1, x_2, x_3, x_4 \geqslant 0, \quad x_3, x_4 \text{ integer.} \end{array}$$

7. Write down a PWL approximation to the following NLP, evaluating $x_1$ at 0, 1, 2 and $x_2$ at 0, 1, 2, 3 and using SOSs.

$$\begin{aligned} \text{maximise} \quad & 5x_2 + x_1^2 - x_2^2 \\ \text{subject to} \quad & 3x_1 + 2x_2 \leqslant 6 \\ & x_1, x_2 \geqslant 0. \end{aligned}$$

Solve the approximate problem. Refine your answer by solving a new PWL approximation, found by dividing the unit square containing the optimal solution of your first problem into four equal quarters.

8. Solve the following TSP.

| | $c_{ij}$ | $j$ = 1 | 2 | 3 | 4 | 5 | 6 |
|---|---|---|---|---|---|---|---|
| $i$ | 1 | – | 11 | 11 | 9 | 6 | 10 |
| | 2 | 7 | – | 8 | 2 | 3 | 5 |
| | 3 | 7 | 3 | – | 3 | 7 | 6 |
| | 4 | 10 | 9 | 4 | – | 4 | 5 |
| | 5 | 16 | 13 | 16 | 13 | – | 9 |
| | 6 | 5 | 10 | 6 | 6 | 8 | – |

CHAPTER 12

# Quadratic Programming

---

## 12.1 THE KUHN–TUCKER CONDITIONS

Quadratic programming deals with problems in which the objective function of an LP is modified by allowing quadratic terms. Such problems can arise out of one approach to multiple objective optimisation (cf. Chapters 9 and 10). Suppose, for example, that $z_1 = x_1 - 5x_2$ and $z_2 = -x_1 + x_2$ are desirable objectives and are subject to the constraints $x_1, x_2 \geqslant 0$ and

$$x_1 - x_2 \leqslant 6$$
$$-x_1 + 5x_2 \leqslant 30.$$

Then $z_1$ is maximised at $(x_1, x_2) = (6, 0)$ giving a maximum value of $z_1 = 6$ and $z_2$ is maximised at (0, 6) with maximal $z_2 = 6$. One suggestion that has been advanced for choosing a 'best' solution is to determine the feasible solution which comes closest to $(z_1, z_2) = (6, 6)$, the **ideal point**, in objective function space. Interpreting 'closest' with the usual definition of distance means minimising

$$(6 - x_1 + 5x_2)^2 + (6 + x_1 - x_2)^2 = 72 + 48x_2 - 12x_1x_2 + 2x_1^2 + 26x_2^2.$$

Dropping a constant term and taking out a factor of two gives problem P1.

$$\begin{aligned} \text{P1:}\quad & \text{maximise} && -24x_2 - x_1^2 - 13x_2^2 + 6x_1x_2 \\ & \text{subject to} && x_1 - x_2 \leqslant 6 \\ &&& -x_1 + 5x_2 \leqslant 30 \\ &&& x_1, x_2 \geqslant 0. \end{aligned}$$

Problem P1 is an example of a **quadratic programming problem** (QP). Another area frequently giving rise to QPs is the fitting of lines and curves to data using a least-squares approach. If certain (linear) constraints on the parameters of the line to be fitted are known, the problem reduces to a QP.

As another example of a QP, we will take problem P2.

P2: maximise $20x_1 - x_2 - 3x_3 - 2x_1^2 - 2x_2^2 - x_3^2 - 3x_1x_2 - 2x_1x_3 - x_2x_3$

subject to

$$\begin{aligned} x_1 - x_2 - 2x_3 &\leqslant 1 \\ -x_1 - x_2 + 4x_3 &\leqslant 1 \\ x_1 + 2x_2 + x_3 &\leqslant 5 \\ x_1, x_2, x_3 &\geqslant 0. \end{aligned}$$

We will write the general form of QP as

QP: maximise $\sum_j c_j x_j - \sum_{j,k} x_j q_{jk} x_k$

subject to $\sum_j a_{ij} x_j \leqslant b_i$ for all $i$

$x_j \geqslant 0$ for all $j$.

Note that the objective function coefficient of $x_j x_k$ for $j \neq k$ in QP is $-q_{jk} - q_{kj}$. If we make the extra assumption that $q_{jk} = q_{kj}$ (symmetry), then this coefficient becomes $-2q_{jk}$. Hence in P2 we have

$$q_{11} = 2, \quad q_{22} = 2, \quad q_{33} = 1, \quad q_{12}\ (= q_{21}) = 1\tfrac{1}{2},$$
$$q_{23}\ (= q_{32}) = \tfrac{1}{2}, \quad q_{31}\ (= q_{13}) = 1.$$

Let us suppose $(\hat{x}_1, \ldots, \hat{x}_n)$ is optimal in QP. We will now see how this optimality can be expressed in LP terms. To do this, let us assume that $(x_1, \ldots, x_n)$ is a feasible solution of QP and observe that, if $\theta > 0$ and $\theta < 1$, then the solution

$$(\hat{x}_1 + \theta(x_1 - \hat{x}_1), \ldots, \hat{x}_n + \theta(x_n - \hat{x}_n)) \tag{12.1}$$

is feasible, since

$$\hat{x}_j + \theta(x_j - \hat{x}_j) = \theta x_j + (1 - \theta)\hat{x}_j \geqslant 0 \text{ (since } x_j \geqslant 0, \hat{x}_j \geqslant 0)$$

and

$$\begin{aligned} \sum_j a_{ij}(\hat{x}_j + \theta(x_j - \hat{x}_j)) &= \theta \sum_j a_{ij} x_j + (1 - \theta) \sum_j a_{ij} \hat{x}_j \\ &\leqslant \theta b_i + (1 - \theta) b_i = b_i. \end{aligned}$$

The objective function value of (12.1), using the symmetry of $q_{jk}$, is

$$\sum_j c_j(\hat{x}_j + \theta(x_j - \hat{x}_j)) - \sum_{j,k} (\hat{x}_j + \theta(x_j - \hat{x}_j)) q_{jk} (\hat{x}_k + \theta(x_k - \hat{x}_k))$$

$$= \sum_j c_j \hat{x}_j - \sum_{j,k} \hat{x}_j q_{jk} \hat{x}_k + \theta \sum_j c_j (x_j - \hat{x}_j) - 2\theta \sum_{j,k} (x_j - \hat{x}_j) q_{jk} \hat{x}_k + \theta^2 K$$

where $K$ is an expression in $x_j$, $\hat{x}_j$, $q_{jk}$ whose precise form need not concern us. The optimality of $(\hat{x}_1, \ldots, \hat{x}_n)$ says that the objective function value of $(\hat{x}_1, \ldots, \hat{x}_n)$ exceeds that of (12.1) and this can be rewritten as

$$\theta \sum_j c_j (x_j - \hat{x}_j) - 2\theta \sum_{j,k} (x_j - \hat{x}_j) q_{jk} \hat{x}_k + \theta^2 K \leqslant 0.$$

Because $\theta > 0$, we can divide by $\theta$ to get

$$\sum_j d_j x_j + \theta K \leqslant \sum_j d_j \hat{x}_j \tag{12.2}$$

where

$$d_j = c_j - 2 \sum_k q_{jk} \hat{x}_k .$$

Thus we must have $\sum_j d_j x_j \leqslant \sum_j d_j \hat{x}_j$, for otherwise (12.2) would be violated for sufficiently small $\theta$. So, $(\hat{x}_1, \ldots, \hat{x}_n)$ is optimal in LQP.

$$\begin{array}{lll} \text{LQP: maximise} & \sum_j d_j x_j & \\ \text{subject to} & \sum_j a_{ij} x_j \leqslant b_i & \text{for all } i, \\ & x_j \geqslant 0 & \text{for all } j. \end{array}$$

Note however, that $d_j$ depends on $(\hat{x}_1, \ldots, \hat{x}_n)$. Indeed it can be seen that for $(\hat{x}_1, \ldots, \hat{x}_n)$ 'reasonably close' to $(x_1, \ldots, x_n)$ the objective function of LQP is an approximation to that of QP, so LQP can be regarded as an LP approximation to QP. It may help to memorise LQP if we observe that

$$d_j = \frac{\partial}{\partial x_j} \left( \sum_j c_j x_j - \sum_{j,k} x_j q_{jk} x_k \right)$$

evaluated at $(\hat{x}_1, \ldots, \hat{x}_n)$. This result was not used in the derivation of LQP but serves as a useful mnemonic.

For example, if we take $(\hat{x}_1, \hat{x}_2) = (15, 9)$ in P1, then

$$d_1 = -2 \times 15 + 6 \times 9 = 24, \quad d_2 = -24 - 26 \times 9 + 6 \times 15 = -168$$

so that LQP becomes P3.

$$\begin{array}{lll} \text{P3: maximise} & 24x_1 - 168x_2 \\ \text{subject to} & x_1 - \phantom{168}x_2 \leqslant 6 \end{array}$$

$$-x_1 + \quad 5x_2 \leqslant 30$$

$$x_1, x_2 \geqslant 0.$$

If (15, 9) were optimal in P1, it would also be optimal in P3. We can test this optimality using the CS conditions. These show that (15, 9) is *not* optimal in P3 and therefore not optimal in P1.

Instead of explicitly writing down LQP and then applying the CS conditions, we could explore these conditions for the general form of LQP. To do this, we first observe that the dual problem of LQP is

$$\begin{aligned} &\text{minimise} && \sum_i b_i \lambda_i && \\ &\text{subject to} && \sum_i \lambda_i a_{ij} \geqslant d_j \left( = c_j - 2 \sum_k q_{jk} \hat{x}_k \right) && \text{for all } j, \\ & && \lambda_i \geqslant 0 && \text{for all } i, \end{aligned}$$

where $\lambda_i$ is the $i$th dual variable. If we write $\mu_j$ for the $j$th dual slack variable, the duality theorem, expressed through the CS conditions, states that, if $(\hat{x}_1, \ldots, \hat{x}_n)$ is optimal in QP, then there exist $\lambda_i, \mu_j$ for $i, j \geqslant 1$ satisfying

$$\begin{aligned} &c_j - 2 \sum_k q_{jk} \hat{x}_k - \sum_i \lambda_i a_{ij} + \mu_j = 0 && && \text{for all } j, \\ &\lambda_i \geqslant 0 \quad \text{for all } i, && \mu_j \geqslant 0 && \text{for all } j, \\ &\lambda_i \hat{s}_i = 0 \quad \text{for all } i, && \mu_j \hat{x}_j = 0 && \text{for all } j, \end{aligned}$$

where $\hat{s}_i$ is the value of the $i$th slack variable in QP. These are called the **Kuhn–Tucker** (KT) conditions for quadratic programming after their discoverers.

For P2, the KT conditions at $(\hat{x}_1, \hat{x}_2, \hat{x}_3) = (3, 0, 1)$ are

$$\begin{aligned} 6 - \lambda_1 + \lambda_2 - \lambda_3 + \mu_1 &= 0 \\ -11 + \lambda_1 + \lambda_2 - 2\lambda_3 + \mu_2 &= 0 \\ -11 + 2\lambda_1 - 4\lambda_2 - \lambda_3 + \mu_3 &= 0 \end{aligned}$$

and $\lambda_i \geqslant 0$, $\mu_j \geqslant 0$, $\lambda_i s_i = 0$, $\mu_j x_j = 0$ for $i, j \geqslant 1$. Since $\hat{x}_1, \hat{x}_3 > 0$ and $(\hat{s}_1, \hat{s}_2, \hat{s}_3) = (0, 0, 1)$, making $\hat{s}_3 > 0$, we have $\lambda_3 = \mu_1 = \mu_3 = 0$ and thus

$$\begin{aligned} -\lambda_1 + \lambda_2 \qquad &= -6 \\ \lambda_1 + \lambda_2 + \mu_2 &= 11 \\ 2\lambda_1 - 4\lambda_2 \qquad &= 11. \end{aligned}$$

These equations have the solution $\lambda_1 = 6\frac{1}{2}$, $\lambda_2 = \frac{1}{2}$, $\mu_2 = 4$ which is non-negative, so the KT conditions are satisfied. We shall discuss in Section 12.3 whether this entitles us to deduce that (3, 0, 1) is optimal in P2.

In our derivation of the KT conditions, we assumed non-negative variables and inequality constraints in QP. The proof that any optimal solution of QP is also optimal in LQP is valid whatever the form of the constraints provided they are linear and are identical in QP and LQP. Thus, the KT conditions can be modified in exactly the same way that the CS conditions were modified in Section 5.4 to allow for free variables and/or equality constraints. This means that, if $x_j$ is a free variable, $\mu_j$ and therefore the requirement $\mu_j \hat{x}_j = 0$ should be omitted from the KT conditions. Also, if the $i$th constraint is an equality $\lambda_i$ is a free variable and, since $\hat{s}_i$ is absent, the requirement $\lambda_i \hat{s}_i = 0$ is omitted.

For example, consider P3.

$$\begin{aligned} \text{P3:} \quad & \text{maximise} \quad -3x_1^2 - x_2^2 - 3x_1x_2 \\ & \text{subject to} \quad x_1 + 2x_2 \leqslant 7 \\ & \qquad\qquad\quad x_1 + x_2 = 3 \\ & \qquad\qquad\quad x_2 \geqslant 0. \end{aligned}$$

Since $x_1$ is a free variable, $(\hat{x}_1, \hat{x}_2) = (-1, 4)$ is feasible and the KT conditions are

$$-6 - \lambda_1 - \lambda_2 \qquad = 0$$

$$-5 - 2\lambda_1 - \lambda_2 + \mu_2 = 0$$

and $\lambda_1 \geqslant 0, \mu_2 \geqslant 0, \lambda_1 \hat{s}_1 = 0, \mu_2 \hat{x}_2 = 0$. Since $\hat{s}_1 = 0$ and $\hat{x}_2 > 0$ then $\mu_2 = 0$ and we must solve

$$\lambda_1 + \lambda_2 = -6$$

$$2\lambda_1 + \lambda_2 = -5.$$

The solution of these equations is $\lambda_1 = 1, \lambda_2 = -7$ and, because $\lambda_2$ is a free variable, this means the KT conditions are satisfied.

## 12.2 USING THE KUHN–TUCKER CONDITIONS

In Section 12.1 we showed that any optimal solution of a QP also satisfies the KT conditions. In this section we will examine, by means of examples, the converse proposition. Is it true that any solution of the KT conditions is optimal in QP? To answer this question, we will examine the problem P4.

$$\begin{aligned} \text{P4:} \quad & \text{maximise} \quad -2x + x^2 \\ & \text{subject to} \quad x \leqslant 3 \quad (x + s = 3, \quad s \geqslant 0) \\ & \qquad\qquad\quad x \geqslant 0. \end{aligned}$$

The KT conditions for P4 are

$$-2 + 2\hat{x} - \lambda_1 + \mu_1 = 0 \tag{12.3}$$

and

$$\lambda_1 \geqslant 0, \quad \lambda_1 s = 0, \quad \mu_1 \geqslant 0, \quad \mu_1 x = 0.$$

For $\hat{x} = 2, \hat{s} = 1$ so that $\mu_1 = \lambda_1 = 0$ and (12.3) is not satisfied. We may conclude that $\hat{x} = 2$ is not optimal in P4. For $\hat{x} = 0, \hat{s} = 3$ so that $\lambda_1 = 0$, and from (12.3), $\mu_1 = 2$. Thus, the KT conditions are satisfied at $\hat{x} = 0$ but this solution is not optimal. To see this we need only observe that the objective function evaluated at $x = 0$ is zero, whereas evaluated at $x = 3$, it is 3. This can also be seen from Fig. 12.1 which displays the graph of $x^2 - 2x$ for $0 \leqslant x \leqslant 3$. It is also clear from the figure that $x = 0$ *is* maximal if we restrict attention to values of $x$ close to zero. We will say that $x = 0$ is a **local maximum**. The formal definition is that $(\hat{x}_1, \ldots, \hat{x}_n)$ is a local maximum of QP if there is some $\epsilon > 0$ making $(\hat{x}_1, \ldots, \hat{x}_n)$ optimal in the problem obtained by adding the constraints

$$\hat{x}_j - \epsilon \leqslant x_j \leqslant \hat{x}_j + \epsilon \qquad \text{for } j \geqslant 1$$

to QP. In some problems, notably where the model imposes restrictions on changes in the variables, it may be appropriate to consider local maxima. However, we will usually be interested in the overall or **global** maximum.

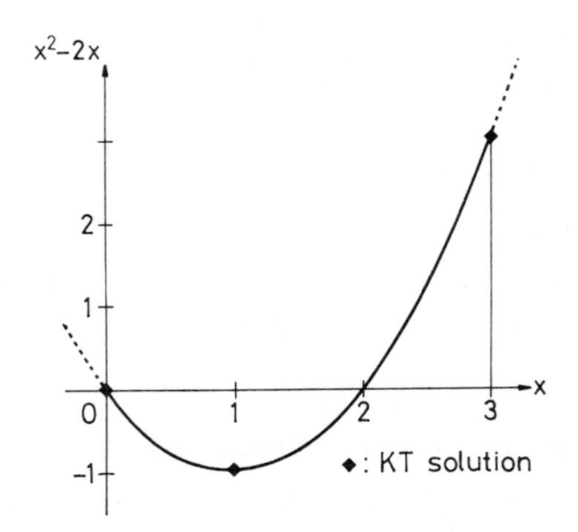


Fig. 12.1 KT solutions of P4.

The proof in Section 12.1, after minor modifications, shows that any local maximum of QP satisfies the KT conditions. This suggests the question: is every solution of the KT conditions a local or global maximum? In P4 we have seen that the local maximum $\hat{x} = 0$ satisfies the KT conditions and we can easily verify the result of Section 12.1, that the global maximum $\hat{x} = 3$ satisfies the KT conditions. Unfortunately, the same is true of $\hat{x} = 1$ (with $\lambda_1 = \mu_1 = 0$) but this is neither a local nor a global maximum. Indeed, it is a global *minimum*. (N.B. The global minimum does not always satisfy the KT conditions. See Section 12.3.)

Our list of types of solution of the KT conditions has now been extended to local and global maximisers and minimisers. Can we stop there or are there yet more types of solution? To answer this question, let us consider problem P5.

$$\text{P5:} \quad \text{maximise} \quad x_1^2 + x_2^2$$
$$\text{subject to} \quad x_1 \leqslant 1 \qquad (x_1 + s_1 = 1, \quad s_1 \geqslant 0).$$

Since both variables are free, the KT conditions are

$$2\hat{x}_1 - \lambda_1 = 0$$
$$2\hat{x}_2 \qquad = 0$$

and $\lambda_1 \geqslant 0, \hat{s}_1\lambda_1 = 0$. These conditions are satisfied at $\hat{x}_1 = 1, \hat{x}_2 = 0$ (with $\lambda_1 = 2$). However, (1, 0) is neither a local (nor a global) maximum nor a minimum since $(1, \alpha)$ has greater objective function value for any $\alpha \neq 0$ and $(1 - \alpha, 0)$ has smaller objective function value for any $\alpha > 0$ $(\alpha < 2)$.

Examples P4 and P5 have shown that for general QPs the force of the KT conditions is essentially negative in that they allow us to conclude that certain solutions (those not satisfying the conditions) are not optimal. We could possibly still use this result constructively by determining all solutions of the KT conditions (usually a finite set) and choosing the solution with largest objective function value. Unfortunately, even this procedure can fail as problem P5 shows. For it is not difficult to see that $(\hat{x}_1, \hat{x}_2) = (1, 0)$ is the *only* solution of the KT conditions and yet, as we have seen, not optimal. This anomaly arises because P5 has no optimal solution; it is unbounded. If we can be sure that QP has an optimal solution, then the procedure of choosing the KT solution with largest objective function value will find that optimum. Sometimes the nature of the underlying model will enable us to conclude that a QP has an optimal solution and it can be shown (see Exercise 2) that, if the constraints are feasible and define a bounded region, then there is an optimal solution. This means that the procedure can be used to solve some small-scale QPs as we shall now illustrate by solving P6.

$$\text{P6:} \quad \text{maximise} \quad -6x_1 + 6x_2 + 3x_1^2 + 3x_2^2 + x_1x_2 \qquad (= z)$$
$$\text{subject to} \quad 3x_1 + 4x_2 \leqslant 12$$
$$x_1, x_2 \geqslant 0.$$

The KT conditions for P6 are

$$-6 + 6\hat{x}_1 + \hat{x}_2 - 3\lambda_1 + \mu_1 = 0 \tag{12.4a}$$

$$6 + \hat{x}_1 + 6\hat{x}_2 - 4\lambda_1 + \mu_2 = 0 \tag{12.4b}$$

and $\lambda_1 \geqslant 0, \hat{s}_1\lambda_1 = 0, \mu_j \geqslant 0, \hat{x}_j\mu_j = 0$ for $j = 1, 2$ where $s_1$ is the slack variable in P6. Each of the variables $x_1, x_2, s_1$ can be positive or zero and this will give us eight cases to consider.

1. $\hat{x}_1 = \hat{x}_2 = \hat{s}_1 = 0$.
   The constraint of P6 is not satisfied.
2. $\hat{x}_1 > 0, \hat{x}_2 = \hat{s}_1 = 0$.
   $\hat{x}_1 = 4$ (from constraint), so $\mu_1 = 0$ ($\hat{x}_1\mu_1 = 0$), $\lambda_1 = 6, \mu_2 = 14$ (from (12.4)). This is a KT solution. $z = 24$.
3. $\hat{x}_1 = 0, \hat{x}_2 > 0, \hat{s}_1 = 0$.
   $\hat{x}_2 = 3, \mu_2 = 0, \lambda_1 = 6, \mu_1 = 0$. A KT solution. $z = 45$ ($> 24$).
4. $\hat{x}_1 = \hat{x}_2 = 0,\ \hat{s}_1 > 0$.
   $z = 0 < 45$ so this solution cannot be optimal.
5. $\hat{x}_1 > 0, \hat{x}_2 > 0, \hat{s}_2 = 0$.
   $\mu_1 = \mu_2 = 0$ and $\hat{x}_1, \hat{x}_2, \lambda_1$ solve:

$$\begin{aligned} 3x_1 + 4x_2 \quad\quad &= 12 \qquad \text{(the constraint)} \\ 6x_1 + x_2 - 3\lambda_1 &= 6 \qquad \text{(from (12.4a))} \\ x_1 + 6x_2 - 4\lambda_1 &= -6 \qquad \text{(from (12.4b))}. \end{aligned}$$

Hence $\hat{x}_1 = 2\frac{2}{3}, \hat{x}_2 = 1, \lambda_1 = 3\frac{2}{3}$. A KT solution. $z = 17 < 45$.

6. $\hat{x}_1 = 0, \hat{x}_2 > 0, \hat{s}_1 > 0$.
   $\lambda_1 = \mu_2 = 0$. (12.4b) would imply $\hat{x}_2 = -1$, so not a KT solution.
7. $\hat{x}_1 > 0, \hat{x}_2 = 0, \hat{s}_1 > 0$.
   $\lambda_1 = \mu_1 = 0$. $\hat{x}_1 = 1, \mu_2 = -7$. Not a KT solution.
8. $\hat{x}_1 > 0, \hat{x}_2 > 0, \hat{s}_1 > 0$.
   $\lambda_1 = \mu_1 = \mu_2 = 0$. (12.4) would imply $\hat{x}_1 = 1\frac{1}{5}, \hat{x}_2 = -1\frac{1}{5}$, so not a KT solution.

The feasible region of P6 is clearly bounded so we can deduce that the solution $(x_1, x_2) = (0, 3)$, in case 3, is optimal, since it has the largest $z$-value of any KT solution. It should be clear that this procedure is only practicable for small problems, so, in the next two sections, we will study a class of problems for which the difficulties described in this section are absent.

## 12.3 CONCAVE QUADRATIC PROGRAMMING

In Section 12.2, we saw that satisfying the KT conditions does not, in itself, guarantee that a solution is optimal, and we examined some of the difficulties that ensue. In this section we will see that, when an extra condition is imposed on the objective function, the KT conditions do guarantee optimality and thus fulfil the same role as the CS conditions for LPs.

We will say that the quadratic form $\sum_{j,k} x_j\, q_{jk}\, x_k$ is **positive semi-definite** (PSD) if, for all $(x_1, \ldots, x_n)$,

$$\sum_{j,k} x_j\, q_{jk}\, x_k \geqslant 0.$$

When this quadratic form is PSD, we will describe the QP as **concave** and we will now show that, if $(\hat{x}_1, \ldots, \hat{x}_n)$ satisfies the KT conditions and QP is concave, then $(\hat{x}_1, \ldots, \hat{x}_n)$ is (globally) optimal. To do this we will take a feasible solution $(x_1, \ldots, x_n)$ of QP and show that its objective function value does not exceed that of $(\hat{x}_1, \ldots, \hat{x}_n)$.

Since $(\hat{x}_1, \ldots, \hat{x}_n)$ satisfies the KT conditions, we have

$$\sum_j x_j \left( c_j - 2 \sum_k q_{jk}\, \hat{x}_k - \sum_i \lambda_i\, a_{ij} + \mu_j \right) = 0$$

and the feasibility of $(x_1, \ldots, x_n)$, together with $\lambda_i \geqslant 0, \mu_j \geqslant 0$, implies

$$\sum_i \lambda_i \sum_j a_{ij}\, x_j \leqslant \sum_i \lambda_i\, b_i\,, \qquad \sum_j x_j\, \mu_j \geqslant 0.$$

Hence, we have

$$\sum_j c_j\, x_j - 2 \sum_{j,k} x_j\, q_{jk}\, \hat{x}_k - \sum_i \lambda_i\, b_i \leqslant 0. \tag{12.5}$$

We also have

$$\sum_j \hat{x}_j \left( c_j - 2 \sum_k q_{jk}\, \hat{x}_k - \sum_i \lambda_i\, a_{ij} + \mu_j \right) = 0$$

and the KT conditions say that $\hat{s}_i \lambda_i = 0$, so that

$$\sum_i \lambda_i \sum_j a_{ij}\, \hat{x}_j = \sum_i \lambda_i \sum_j (a_{ij}\, \hat{x}_j + \hat{s}_i) = \sum_i \lambda_i\, b_i.$$

Since $\hat{x}_j\, \mu_j = 0$, we can deduce that

$$\sum_j c_j\, \hat{x}_j - 2 \sum_{j,k} \hat{x}_j\, q_{jk}\, \hat{x}_k - \sum_i \lambda_i\, b_i = 0. \tag{12.6}$$

The concavity of QP implies that

$$\sum_{j,k} (x_j - \hat{x}_j)\, q_{jk}\, (x_k - \hat{x}_k) \geqslant 0,$$

which can be restated as

$$- \sum_{j,k} x_j\, q_{jk}\, x_k + 2 \sum_{j,k} x_j\, q_{jk}\, \hat{x}_k - \sum_{j,k} \hat{x}_j\, q_{jk}\, \hat{x}_k \leqslant 0. \tag{12.7}$$

Adding (12.5) to (12.7) and subtracting (12.6) yields

$$\sum_j c_j x_j - \sum_{j,k} x_j q_{jk} x_k \leqslant \sum_j c_j \hat{x}_j - \sum_{j,k} \hat{x}_j q_{jk} \hat{x}_k$$

which establishes the desired optimality of $(\hat{x}_1, \ldots, \hat{x}_n)$.

We have shown that, in the concave case, we have a complete test for optimality by checking whether a putative optimal solution satisfies the KT conditions. For example, the KT conditions for problem P1 are

$$-2\hat{x}_1 + 6\hat{x}_2 - \lambda_1 + \lambda_2 + \mu_1 = 0$$

$$-24 + 6\hat{x}_1 - 26\hat{x}_2 + \lambda_1 - 5\lambda_2 + \mu_2 = 0$$

and $\lambda_i \geqslant 0$, $\hat{s}_i \lambda_i = 0$, $\mu_j \geqslant 0$, $\hat{x}_j \mu_j = 0$ for $i, j = 1, 2$. These conditions are satisfied at $(\hat{x}_1, \hat{x}_2) = (0, 0)$, with $\lambda_1 = \lambda_2 = \mu_1 = 0$, $\mu_2 = 24$. Furthermore, the quadratic form in the objective function of P1 can be written

$$x_1^2 + 13x_2^2 - 6x_1x_2 = (x_1 - 3x_2)^2 + 4x_2^2 \geqslant 0$$

showing that P1 is concave and, therefore, that (0, 0) is optimal.

We were able to recognise the concavity of P1 by re-expressing the quadratic terms in the objective function. Very often, the nature of the model underlying the QP enables us to deduce the concavity. This is true for least-squares applications in statistics and for the approach to multiple-objective problems outlined in Section 12.1. To justify the latter statement, we need only observe that, if $\sum_j c_{Kj} x_j$ is the $K$th objective function and $\mu_K$ is its maximal value (the $K$th component of the ideal point), then our objective is to maximise

$$-\sum_K \left(\mu_K - \sum_j c_{Kj} x_j\right)^2.$$

Consequently, in the standard notation for QP,

$$\sum_{j,k} x_j q_{jk} x_k = \sum_K \left(\sum_j c_{Kj} x_j\right)^2 \geqslant 0,$$

so the quadratic form is PSD.

We are still left with the problem of deciding whether an arbitrary quadratic form is PSD and we shall now explain how the question can be settled by generalising the trick used on P1. Consider

$$2x_1^2 + 2x_2^2 + x_3^2 + 3x_1x_2 + 2x_1x_3 + x_2x_3.$$

The terms involving $x_1$ in this expression are

$$2x_1^2 + 3x_1x_2 + 2x_1x_3 = 2\,(x_1^2 + 1\tfrac{1}{2}x_1x_2 + x_1x_3).$$

Hence, if we put $y_1 = x_1 + \frac{3}{4}x_2 + \frac{1}{2}x_3$ so that $y_1^2 = x_1^2 + 1\frac{1}{2}x_1x_2 + x_1x_3 + \frac{9}{16}x_2^2 + \frac{1}{4}x_3^2 + \frac{3}{4}x_2x_3$, the quadratic form can be rewritten as

$$2y_1^2 + \tfrac{7}{8}x_2^2 + \tfrac{1}{2}x_3^2 - \tfrac{1}{2}x_2x_3.$$

Continuing the process by putting $y_2 = x_2 - \frac{2}{7}x_3$, the quadratic form becomes

$$2y_1^2 + \tfrac{7}{8}y_2^2 + \tfrac{3}{7}x_3^2$$

and this is clearly PSD. Since this quadratic form is just the quadratic part of the objective function of P2, we have shown that P2 is concave and, therefore, that $(x_1, x_2, x_3) = (3, 0, 1)$, which satisfies the KT conditions as we showed in Section 12.1, is optimal.

In general, assuming $q_{11} > 0$, we put

$$y_1 = \left(\sum_k q_{1k}x_k\right)/q_{11} = x_1 + \sum_{k\geq 2}(q_{1k}x_k)/q_{11}$$

and observe that

$$\sum_{j,k} x_j q_{jk} x_k = \delta_1 y_1^2 + \sum_{j,k\geq 2} x_j q_{jk}^* x_k,$$

where $\delta_1 = y_{11}$ and

$$q_{jk}^* = q_{jk} - (q_{j1}q_{1k}/q_{11}). \tag{12.8}$$

This process can then be applied to the remaining terms and repeated as often as necessary. If it fails at any stage (because $q_{11} \leqslant 0$), we can deduce that the quadratic form is not PSD as we shall show below.

For example,

$$x_1^2 + 5x_2^2 + 7x_3^2 + 6x_1x_2 + 4x_1x_3 + 7x_2x_3$$

can be rewritten as

$$y_1^2 - 4x_2^2 + 3x_3^2 - 5x_2x_3$$

where $y_1 = x_1 + 3x_2 + 2x_3$. Here $x_2^2$ has a negative coefficient so the process cannot be applied, but if we put $x_2 = 1, x_3 = 0$ and $x_1 = -3$ (to make $y_1 = 0$), we see that the value of the quadratic form is $-4$. Hence it is not PSD.

In general, if we have $q'_{rr} < 0$, where $r \geqslant 2$, and

$$\sum_{j=1}^{r-1} \delta_j y_j^2 + \sum_{j,k\geq r} x_j q'_{jk} x_k \tag{12.9}$$

is derived from the original quadratic form by the process described above, then putting $x_r = 1, x_j = 0$ for $j \geqslant r + 1$ and choosing $x_{r-1}, \ldots, x_1$ to make $y_{r-1} = \ldots = y_1 = 0$ (which is always possible), expression (12.9) has the value $q'_{rr}$ ($< 0$) so the quadratic form is not PSD. A similar argument shows that, if $q'_{jj} < 0$ for any $j \geqslant r$, the quadratic form is not PSD. So, if we had

$$y_1^2 + 3x_2^2 - 2x_3^2 + 7x_2x_3,$$

for example, we could deduce that the quadratic form is not PSD.

We might also find $q'_{rr} = 0$ in (12.9). This poses no problems if $x_r$ does not occur at all in this expression, that is $q'_{jr} = q'_{rj} = 0$ for $j \geqslant r$. We simply go on to look at $x_{r+1}$. For example, if the process leads to

$$9y_1^2 + 0x_2^2 + 2x_3^2 + 6x_4^2 + 3x_3x_4$$

we ignore $x_2$ and put $y_3 = x_3 + \frac{3}{4}x_4$, to deduce that the quadratic form is PSD.

If $q'_{rr} = 0$ and $q'_{rt} \neq 0$ for some $t > r$, we can deduce that the quadratic form is not PSD. For, if we put $x_r = -1 - q'_{tt}$, $x_t = 2q'_{rt}$, $x_j = 0$ if $j \geqslant r$ otherwise, and choose $x_{r-1}, \ldots, x_1$ to make $y_{r-1} = \ldots = y_1 = 0$, then (12.9) has value

$$q'_{tt}\,x_t^2 + 2q'_{rt}\,x_r\,x_t = -4(q'_{rt})^2 < 0.$$

Thus the quadratic form is not PSD. For example, if the process had led to

$$9y_1^2 + 0x_2^2 + 4x_3^2 + 6x_4^2 + 5x_2x_4 + 3x_3x_4,$$

then putting $x_2 = -1 - 6 = -7, x_4 = 2 \times 2\frac{1}{2} = 5$ $(r = 2, t = 4)$ and $x_3 = y_1 = 0$ gives a value of $-25$.

We have now covered all cases and have seen that the process either expresses the quadratic form as a sum of squares which is PSD or allows us to deduce that it is not PSD. It is interesting to compare formula (12.8) with the transformation formulae for pivoting. The similarity which is observed suggests that the operations performed algebraically above can be carried out using (slightly modified) pivoting in a tableau. The suggestion is pursued further in Exercise 4.

## 12.4 SOLVING THE CONCAVE PROBLEM

In Section 12.3 we saw that any feasible solution of the KT conditions for a concave QP is optimal. In Section 12.2 we suggested a systematic way of searching for KT solutions but applicable only to small problems. In this section we will show how KT solutions of concave QPs may be found by pivoting, enabling much larger problems to be solved. We will illustrate the procedure by solving problem P7.

P7: maximise $8x_1 + 4x_2 - x_1^2 + x_1x_2 - x_2^2$

subject to

$$x_1 + x_2 \leqslant 4$$

$$x_1 - x_2 \geqslant 1$$

$$x_1, x_2 \geqslant 0.$$

Using the process described in Section 12.3 we can verify that P7 is concave and deduce that $(\hat{x}_1, \hat{x}_2)$ is optimal if it satisfies the KT conditions:

$$-2x_1 + x_2 - \lambda_1 + \lambda_2 + \mu_1 = -8$$

$$x_1 - 2x_2 - \lambda_1 - \lambda_2 + \mu_2 = -4$$

and $\lambda_i \geqslant 0, s_i\lambda_i = 0, \mu_j \geqslant 0, x_j\mu_j = 0$ for $i, j = 1, 2$, and is feasible:

$$\begin{aligned} x_1 + x_2 \qquad + s_1 &= \ \ 4 \\ -x_1 + x_2 \qquad + s_2 &= -1 \end{aligned}$$

and $x_j$, $s_i \geqslant 0$ for $i, j = 1, 2$. These four equations have been written in tableau form in P7/T1 together with an extra ($\nu-$) column in which every entry is $-1$. We will explain subsequently how this column is used to obtain an initial feasible solution.

| P7 | $x_1$ | $x_2$ | $\lambda_1$ | $\lambda_2$ | $\nu$ | T1 |
|---|---|---|---|---|---|---|
| $\mu_1$ | $-2$ | 1 | $-1$ | 1 | ($-1$) | $-8$ |
| $\mu_2$ | 1 | $-2$ | $-1$ | $-1$ | $-1$ | $-4$ |
| $s_1$ | 1 | 1 | 0 | 0 | $-1$ | 4 |
| $s_2$ | $-1$ | 1 | 0 | 0 | $-1$ | $-1$ |

To assist our description of the method, we will extend the terminology used for duality in LPs by describing $(x_j, \mu_j)$ and $(\lambda_i, s_i)$ as **complementary variables**. Any tableau in which one variable in each complementary pair is basic and the other non-basic will be called a **complementary tableau**. We will also encounter tableaux which fail to meet these requirements only to the extent of having a single pair of complementary non-basic variables (the complementary variable of any basic variable being non-basic). We shall refer to such tableaux as **almost-complementary** (AC).

P7/T1 is a complementary tableau but is not primal feasible. However, this can be remedied by pivoting in the $\nu$-column as indicated to give P7/T2, an AC tableau.

| P7 | $x_1$ | $x_2$ | $\lambda_1$ | $\lambda_2$ | $\mu_1$ | T2 |
|---|---|---|---|---|---|---|
| $\nu$ | 2 | $-1$ | 1 | $-1$ | $-1$ | 8 |
| $\mu_2$ | (3) | $-3$ | 0 | $-2$ | $-1$ | 4 |
| $s_1$ | 3 | 0 | 1 | $-1$ | $-1$ | 12 |
| $s_2$ | 1 | 0 | 1 | $-1$ | $-1$ | 7 |

The pivot row in P7/T1 was chosen to be a row with most negative resource column entry. This rule will always ensure primal feasibility on pivoting. For, if $\alpha_{i0}$ denotes the resource column entry in row $i$, and we pivot in row $r$, where

$\alpha_{ro} = \min\{\alpha_{io}\} < 0$, then $\alpha'_{ro} = \alpha_{ro}/(-1) > 0$ and

$$\alpha'_{io} = \alpha_{io} - (-1)\,\alpha_{ro}/(-1) = \alpha_{io} - \alpha_{ro} \geqslant 0 \quad \text{for } i \neq r.$$

P7/T2 is a primal-feasible, AC tableau. All that prevents the associated BFS from being a feasible KT solution is that $\nu > 0$. If we can find a primal-feasible, complementary tableau ($\nu$ non-basic will ensure this), then the associated BFS satisfies the KT conditions and is therefore optimal. We will proceed by generating a sequence of primal-feasible, AC tableaux by pivoting, hoping that in one of them $\nu$ leaves the basis. In P7/T2, both $x_1$ and $\mu_1$ are non-basic and we could make one of them basic without violating the requirement $x_1\mu_1 = 0$. Using the PRS rule to maintain primal feasibility, the pivot in the $x_1$-column is in the $\mu_2$-row. There is no positive pivot in the $\mu_1$-column, so we make the indicated pivot to arrive at P7/T3.

| P7 | $\mu_2$ | $x_2$ | $\lambda_1$ | $\lambda_2$ | $\mu_1$ | T3 |
|---|---|---|---|---|---|---|
| $\nu$ | $-\frac{2}{3}$ | 1 | 1 | $\frac{1}{3}$ | $-\frac{1}{3}$ | $5\frac{1}{3}$ |
| $x_1$ | $\frac{1}{3}$ | $-1$ | 0 | $-\frac{2}{3}$ | $-\frac{1}{3}$ | $1\frac{1}{3}$ |
| $s_1$ | $-1$ | (3) | 1 | 1 | 0 | 8 |
| $s_2$ | $-\frac{1}{3}$ | 1 | 1 | $-\frac{1}{3}$ | $-\frac{2}{3}$ | $5\frac{2}{3}$ |

In P7/T3 we could pivot in the $\mu_2$- or $x_2$-column. But, pivoting in the $\mu_2$-column would return us to P7/T2, so we pivot in the $x_2$-column. In general, we pivot in the column containing the complementary variable of the one that has just left the basis (at the previous pivot). This ensures almost-complementarity of successive tableaux. Making the indicated pivot gives tableau P7/T4 and one further iteration gives P7/T5.

| P7 | $\mu_2$ | $s_1$ | $\lambda_1$ | $\lambda_2$ | $\mu_1$ | T4 |
|---|---|---|---|---|---|---|
| $\nu$ | $-\frac{1}{3}$ | $-\frac{1}{3}$ | ($\frac{2}{3}$) | 0 | $-\frac{1}{3}$ | $2\frac{2}{3}$ |
| $x_1$ | 0 | $\frac{1}{3}$ | $\frac{1}{3}$ | $-\frac{1}{3}$ | $-\frac{1}{3}$ | 4 |
| $x_2$ | $-\frac{1}{3}$ | $\frac{1}{3}$ | $\frac{1}{3}$ | $\frac{1}{3}$ | 0 | $2\frac{2}{3}$ |
| $s_2$ | 0 | $-\frac{1}{3}$ | $\frac{2}{3}$ | $-\frac{2}{3}$ | $-\frac{2}{3}$ | 3 |

P7/T5 is complementary because $\nu$ is non-basic and therefore $(\hat{x}_1, \hat{x}_2) = (2\frac{2}{3}, 1\frac{1}{3})$ satisfies the KT conditions, with $\lambda_1 = 4, \mu_1 = \mu_2 = \lambda_2 = 0$. This can readily be verified directly.

| P7 | $\mu_2$ | $s_1$ | $\nu$ | $\lambda_2$ | $\mu_1$ | T5 |
|---|---|---|---|---|---|---|
| $\lambda_1$ | $-\frac{1}{2}$ | $-\frac{1}{2}$ | $\frac{3}{2}$ | 0 | $-\frac{1}{2}$ | 4 |
| $x_1$ | $\frac{1}{6}$ | $\frac{1}{2}$ | $-\frac{1}{2}$ | $-\frac{1}{3}$ | $-\frac{1}{6}$ | $2\frac{2}{3}$ |
| $x_2$ | $-\frac{1}{6}$ | $\frac{1}{2}$ | $-\frac{1}{2}$ | $\frac{1}{3}$ | $\frac{1}{6}$ | $1\frac{1}{3}$ |
| $s_2$ | $\frac{1}{3}$ | 0 | $-1$ | $-\frac{2}{3}$ | $-\frac{1}{3}$ | $\frac{1}{3}$ |

We will show in Section 12.5 that the method will 'solve' any concave QP in the sense that it will find an optimal solution or indicate that there is no optimal solution. For the remainder of this section, we will see how the method can be modified to cope with equality constraints by solving problem P8.

$$\begin{aligned} \text{P8:}\quad & \text{maximise} && -x_1^2 - x_2^2 \\ & \text{subject to} && x_1 + x_2 \geqslant 2 \\ &&& x_1 - x_2 = 1 \\ &&& x_1, x_2 \geqslant 0. \end{aligned}$$

The KT conditions and feasibility requirements for P8 are

$$\begin{aligned} -2x_1 + \lambda_1 - \lambda_2 + \mu_1 &= 0 \\ -2x_2 + \lambda_1 + \lambda_2 + \mu_2 &= 0 \\ -x_1 - x_2 + s_1 &= -2 \\ x_1 - x_2 &= 1 \end{aligned}$$

and $x_1\mu_1 = x_2\mu_2 = s_1\lambda_1 = 0$ with all variables except $\lambda_2$ non-negative. These equations are represented in the tableau P8/T1 in which we have also introduced an artificial variaable, $a_2$ as basic variable in the equality constraint.

| P8 | $x_1$ | $x_2$ | $\lambda_1$ | $\lambda_2$ | T1 |
|---|---|---|---|---|---|
| $\mu_1$ | $-2$ | 0 | 1 | $-1$ | 0 |
| $\mu_2$ | 0 | $-2$ | 1 | 1 | 0 |
| $s_1$ | $-1$ | $-1$ | 0 | 0 | $-2$ |
| $a_2$ | (1) | $-1$ | 0 | 0 | 1 |

We wish to make $a_2$ non-basic, but there is no need to introduce an infeasibility or any other objective function. Since we will not impose primal feasibility in this part of the method, any non-zero pivot in the $a_2$-row will do.

Making the indicated pivot in P8/T1 gives P8/T2 from which the $a_2$-column has been dropped, as usual.

| P8 | $x_2$ | $\lambda_1$ | $\lambda_2$ | T2 |
|---|---|---|---|---|
| $\mu_1$ | −2 | 1 | (−1) | 2 |
| $\mu_2$ | −2 | 1 | 1 | 0 |
| $s_1$ | −2 | 0 | 0 | −1 |
| $x_1$ | −1 | 0 | 0 | 1 |

Tableau P8/T2 is not complementary; $\mu_1$ and $x_1$ are both basic. However, $\lambda_2$ has no complementary variable so pivoting in the $\mu_1$-row and $\lambda_2$-column restores complementarity. So $\lambda_2$, a free variable, becomes basic. When a free variable is basic in any tableau, its row will never be the pivot row, since there is no requirement that the variable be non-negative. Consequently, we can drop the row. Pivoting as indicated in P8/T2, dropping the resulting $\lambda_2$-row and adding a $\nu$-column gives tableau P8/T3.

| P8 | $x_2$ | $\lambda_1$ | $\mu_1$ | $\nu$ | T3 |
|---|---|---|---|---|---|
| $\mu_2$ | −4 | 2 | 1 | −1 | 2 |
| $s_1$ | −2 | 0 | 0 | (−1) | −1 |
| $x_1$ | −1 | 0 | 0 | −1 | 1 |

In general, if $a_i$ is an artificial variable corresponding to the $i$th constraint (an equality), $a_i$ is pivoted out of the basis. Immediately afterwards, we pivot in the $\lambda_i$-column and the row containing the complementary variable of the one which became basic at the previous pivot. This gives a complementary tableau. We continue performing pairs of pivots and dropping rows and columns until no artificial variables are left in the tableau. The only occurrence which could prevent us applying this method is a zero pivot (negative pivots are acceptable) and it can be shown that this cannot occur (Exercise 8). The result is a complementary tableau to which a $\nu$-column can be added and we then proceed exactly as before.

| P8 | $x_2$ | $\lambda_1$ | $\mu_1$ | $s_1$ | T4 |
|---|---|---|---|---|---|
| $\mu_2$ | −2 | (2) | 1 | −1 | 3 |
| $\nu$ | 2 | 0 | 0 | −1 | 1 |
| $x_1$ | 1 | 0 | 0 | −1 | 2 |

Pivoting as indicated in P8/T3 gives P8/T4. Two further pivots lead to P8/T5, which is complementary. Hence $(\hat{x}_1, \hat{x}_2) = (1\frac{1}{2}, \frac{1}{2})$ is optimal in P8.

| P8 | $\nu$ | $\mu_2$ | $\mu_1$ | $s_1$ | T5 |
|---|---|---|---|---|---|
| $\lambda_1$ | $\frac{1}{2}$ | $\frac{1}{2}$ | $\frac{1}{2}$ | $-1$ | 2 |
| $x_2$ | $\frac{1}{2}$ | 0 | 0 | $-\frac{1}{2}$ | $\frac{1}{2}$ |
| $x_1$ | $-\frac{1}{2}$ | 0 | 0 | $-\frac{1}{2}$ | $1\frac{1}{2}$ |

## 12.5 FINITENESS OF THE METHOD

In this section we will show that for any concave QP, the method described in Section 12.4 either finds an optimal solution or allows us to deduce that there is no such solution. Throughout this section we shall assume non-degeneracy. Degeneracy could be handled by techniques such as that described in Section 3.6. Moreover, degeneracy has not led to serious computational problems and can be ignored in practice.

Our first task will be to show that cycling cannot occur. Since there is no objective function to increase we cannot use the same argument as for the simplex method (Section 3.4). The proof is based on the observation that every tableau of the method, except the initial one ($\nu$ non-basic) and any final, complementary one (again, $\nu$ non-basic), is AC. Consequently, there are exactly two tableaux that can be reached from any tableau by pivoting, if we wish to preserve zero products of complementary variables and primal feasibility. One of this pair of tableaux will be the tableau for the previous iteration. The method prescribes that we pivot to reach the other. No tableau can repeat under this procedure for, if a pivot from T$i$ led to the already encountered tableau T$j$ we would have to have $j \leqslant i-2$ (since we never pivot to the preceding tableau). But this would mean we could reach three tableaux: T$(j-1)$, T$(j+1)$ and T$i$ from T$j$, contradicting our previous observation unless $j = 1$ (initial tableau). But the PRS rule would prevent us pivoting from T$i$ to non-primal-feasible T1.

Since cycling cannot occur and there are only finitely many tableaux, the method must terminate. It can only terminate if $\nu$ becomes non-basic, giving an optimal solution to QP, or if there is no pivot because every element in the pivot column is non-positive. For example, consider problem P9.

$$\begin{aligned} \text{P9:}\quad &\text{maximise,} && 3x_1 - x_2 - 4x_1^2 + 4x_1x_2 - x_2^2 \\ &\text{subject to} && 3x_1 - 2x_2 \leqslant 2 \\ &&& x_1, x_2 \geqslant 0. \end{aligned}$$

The initial tableau is P9/T1 and two pivots lead to P9/T2.

| P9 | $x_1$ | $x_2$ | $\lambda_1$ | $\nu$ | T1 |
|---|---|---|---|---|---|
| $\mu_1$ | $-8$ | 4 | $-3$ | $(-1)$ | $-3$ |
| $\mu_2$ | 4 | $-2$ | 2 | $-1$ | 1 |
| $s_1$ | 3 | $-2$ | 0 | $-1$ | 2 |

| P9 | $\mu_2$ | $x_2$ | $\lambda_1$ | $\mu_1$ | T2 |
|---|---|---|---|---|---|
| $\nu$ | $-\frac{2}{3}$ | 0 | $-\frac{1}{3}$ | $-\frac{1}{3}$ | $\frac{1}{3}$ |
| $x_1$ | $\frac{1}{12}$ | $-\frac{1}{2}$ | $\frac{5}{12}$ | $-\frac{1}{12}$ | $\frac{1}{3}$ |
| $s_1$ | $-\frac{11}{12}$ | $-\frac{1}{2}$ | $-\frac{19}{12}$ | $-\frac{1}{12}$ | $1\frac{1}{3}$ |

In P9/T2, we wish to make $x_2$ basic but there is no positive element in the pivot column. In addition, the first row of P9/T2 shows that

$$\nu = \tfrac{1}{3} + \tfrac{2}{3}\mu_2 + \tfrac{1}{3}\lambda_1 + \tfrac{1}{3}\mu_1 \geqslant \tfrac{1}{3}$$

in any feasible solution. Hence there is no solution with $\nu = 0$, the KT conditions have no solution and, since it is concave, P9 has no optimal solution. We will now see that whenever the pivot column lacks a pivot, the $\nu$-row also has non-positive entries, except in the resource column, showing that $\nu = 0$ is infeasible. Hence the method always finds an optimal solution if there is one.

In order to prove this we need a preliminary result about quadratic forms. Suppose that QP is concave and $\sum_{j,k} \bar{x}_j q_{jk} \bar{x}_k = 0$, then $\sum_k q_{jk} \bar{x}_k = 0$ for all $j$. To show this, we observe that for any $(x_1, \ldots, x_n)$ and $\theta > 0$,

$$\sum_{j,k} (\bar{x}_j + \theta x_j) q_{jk} (\bar{x}_k + \theta x_k) \geqslant 0.$$

This can be rewritten, after dividing by $\theta$, as

$$2 \sum_{j,k} x_j q_{jk} \bar{x}_k + \theta \sum_{j,k} x_j q_{jk} x_k \geqslant 0 \qquad \text{for all } \theta > 0.$$

Hence, $\sum_{j,k} x_j q_{jk} \bar{x}_k \geqslant 0$. Putting $x_j = -\sum_k q_{jk} \bar{x}_k$, gives

$$\sum_j \left( \sum_k q_{jk} \bar{x}_k \right)^2 \leqslant 0$$

and this can only be true if every term in the sum is zero, as claimed.

Returning to our solution method for QPs, let us suppose we have an AC tableau T* in which, according to the method, we must pivot in column $k$, but (in standard notation) $\alpha_{ik} \leqslant 0$ for all $i$. If $\nu$ is the basic variable in row $r$, we must show that $\alpha_{rj} \leqslant 0$ for all $j$. We will do this by observing that the solution

$$(*): x_{N_k} = 1, \quad x_{N_j} = 0 \quad \text{for } j \neq k, \quad x_{B_i} = -\alpha_{ik} \geqslant 0 \quad \text{for } i \geqslant 1$$

is non-negative, satisfies the equations defined by the tableau T$^*$, but with RHSs zero and, because the tableau is AC, the product of complementary variables is zero. Now a resource column of zeros remains a resource colum of zeros after pivoting. Thus, this solution, which we write $(\bar{x}_j, \bar{\lambda}_i, \bar{\mu}_j, \bar{s}_i, \bar{\nu}) \geqslant 0$ in terms of the original variables, must satisfy the equations of the initial tableau with a resource column of zeros. In the case of P9/T2 we have $\bar{x}_2 = 1, \bar{x}_1 = \frac{1}{2}, \bar{s}_1 = \frac{1}{2}$, $\bar{\nu} = \bar{\mu}_2 = \bar{\lambda}_1 = \bar{\mu}_1 = 0$ and it can be checked that these satisfy the equations of P9/T1 with zero RHSs.

Written explicitly, these equations are

$$-2\sum_k q_{jk}\,\bar{x}_k - \sum_i \bar{\lambda}_i\, a_{ij} + \bar{\mu}_j - \bar{\nu} = 0 \quad \text{for all } j, \tag{12.10a}$$

$$\sum_j a_{ij}\,\bar{x}_j + \bar{s}_i - \bar{\nu} = 0 \quad \text{for all } i. \tag{12.10b}$$

Multiplying (12.10a) by $\bar{x}_j$, (12.10b) by $\bar{\lambda}_i$, summing and using $\bar{x}_j\bar{\mu}_j = \bar{s}_i\,\bar{\lambda}_i = 0$ for all $i, j$, we obtain

$$-2\sum_{j,k} \bar{x}_j\, q_{jk}\,\bar{x}_k = \bar{\nu}\left(\sum_j \bar{x}_j + \sum_i \bar{\lambda}_i\right) = \bar{\nu}\bar{\omega}, \text{ say.}$$

Since the left-hand side of this equation is non-positive and the right-hand side non-negative, both sides must equal zero. Furthermore, $\bar{x}_j = 0$ and $\bar{\lambda}_i = 0$ for all $i$ and $j$ would imply that these were the non-basic variables not in the pivot column in tableau T$^*$. But this would mean we were trying to pivot back to the initial tableau, which is not allowed in the method. So $\bar{\omega} > 0$, and thus $\bar{\nu} = 0$ and, by the result proved above, $\sum_k q_{jk}\,\bar{x}_k =$ for all $j$.

Now the initial tableau with resource column of zeros corresponds to the equations

$$-2\sum_k q_{jk}\, x_k - \sum_i \lambda_i\, a_{ij} + \mu_j - \nu = 0 \tag{12.11a}$$

$$\sum_j a_{ij}\, x_j + s_i - \nu = 0. \tag{12.11b}$$

Let us investigate the effect of introducing an objective function

$$z = \sum_j (\bar{\mu}_j\, x_j + \bar{x}_j\, \mu_j) + \sum_i (\bar{\lambda}_i\, s_i + \bar{s}_i\, \lambda_i) - \bar{\omega}\nu.$$

Multiplying (12.11a) by $\bar{x}_j$, (12.11b) by $\bar{\lambda}_i$, summing and using (12.10) and $\bar{\nu} =$

$\sum_k q_{jk} \bar{x}_k = 0$ gives $z = 0$ for any feasible solution of (12.11). This means we have an objective row of zeros in the initial tableau.

In P9, this objective function is

$$z = \tfrac{1}{2}\mu_1 + \mu_2 + \tfrac{1}{2}\lambda_1 - 1\tfrac{1}{2}\nu$$

and it is readily checked that if this is used in P9/T1, an objective row of zeros is obtained.

Consequently, we have an objective row of zeros in $T^*$ also. We can then apply formula (3.6) to $T^*$, noting that the coefficient of $x_{B_r}$ (i.e. $\nu$) in the objective function is $-\bar{\omega}$. Furthermore, the coefficient of $x_{B_i}$ for $i \neq r$ is zero, since it is the value in (*) of the complementary variable of $x_{B_i}$, which is $x_{N_j}$ for some $j \neq k$. Thus (3.6) gives

$$-\bar{\omega}\alpha_{rj} - c_{N_j} = 0 \qquad \text{for } j \neq k.$$

Since all objective function coefficients are non-negative and $\bar{\omega} > 0$, we deduce that $\alpha_{rj} \leqslant 0$ for all $j$ ($\alpha_{rk} \leqslant 0$, by assumption).

The argument extends to the case when QP contains equality constraints provided the objective function $z$ is defined to only include $\lambda_i$ when the $i$th constraint is an inequality.

## EXERCISES

1. Solve the following problems by finding all the KT solutions. Verify (graphically for problem (ii)) that, in each case, the feasible region is bounded.

(i) maximise $-x_1 + x_1^2 - x_1x_2 + x_2^2$

subject to $x_1 + 2x_2 \leqslant 3$

$x_1, x_2 \geqslant 0.$

(ii) maximise $x_1 + 6x_2 + x_1^2 + x_2^2$

subject to $x_1 + x_2 \leqslant 6$

$x_1 + 4x_2 \geqslant 8$

$4x_1 + x_2 \geqslant 5.$

(iii) maximise $-x_1 + 2x_3 - x_1^2 + 2x_2^2 - 3x_3^2 - 2x_1x_2 + 4x_2x_3$

subject to $2x_1 + x_2 + x_3 = 6$

$x_1 + x_2 + x_3 \leqslant 5$

$x_1, x_2, x_3 \geqslant 0.$

In case (iii), first show that $x_1 > 0$ in any feasible solution.

2. (Assumes a knowledge of real analysis.)

Show that, if the feasible region of a QP is bounded and non-empty, the QP has an optimal solution. Deduce that any QP is either infeasible, unbounded or has an optimal solution.

3. Which of the following quadratic forms is PSD?

(i) $2x_1^2 + 5x_2^2 + 7x_3^2 - 6x_1x_2 + 6x_1x_3 - 10x_2x_3$.
(ii) $2x_1^2 + 8x_2^2 + 3x_3^2 + 8x_1x_2 + 4x_1x_3 + 10x_2x_3$.
(iii) $x_1^2 + 16x_2^2 + 9x_3^2 + 8x_1x_2 + 6x_1x_3 - 2x_2x_3$.

4. Show how the method you used to answer Exercise 3 can be performed by pivoting in a tableau.

5. If $\sum_{j,k} x_j q_{jk} x_i > 0$ for all $(x_1, \ldots, x_n)$ except $x_1 = \ldots = x_n = 0$, then the quadratic form is said to be **positive definite** (PD). Show that if the quadratic terms of a feasible QP are PD, then the problem has a unique optimal solution. Which of the quadratic forms in Exercise 3 is PD?

6. For each of the following problems and trial solutions, test the trial solution for optimality, carefully justifying your conclusion.

(i) maximise $-7x_1 + x_2 - x_3 - 10x_1^2 - 3x_2^2 - 3x_3^2 - 10x_1x_2 - 8x_1x_3 - 4x_2x_3$

subject to
$$\begin{aligned} 7x_1 + 3x_2 &\geq 11 \\ 32x_1 + 8x_2 + 4x_3 &\geq 53 \\ 5x_1 + x_2 + 2x_3 &\geq 9 \\ x_1 + x_2 &\geq 0 \\ x_1 - x_2 &\geq 0 \\ x_3 &\geq 0. \end{aligned}$$

Trial solution: $(2, -1, 0)$

(ii) minimise $3x_1^2 + 16x_2^2 + 9x_3^2 + 8x_1x_2 + 6x_1x_3 - 2x_2x_3$

subject to
$$\begin{aligned} x_1 + 2x_2 - 30x_3 &\leq -29 \\ 9x_1 + 3x_2 - 2x_3 &\geq 7 \\ x_1 + 3x_2 + 5x_3 &\leq 7 \\ x_1, x_2, x_3 &\geq 0. \end{aligned}$$

Trial solution: $(1, 0, 1)$.

(iii) maximise $7x_1 - x_2 + 9x_3 - 2x_1^2 - 5x_2^2 - 7x_3^2 + 6x_1x_2 - 6x_1x_3 + 10x_2x_3$

subject to
$$6x_1 + x_2 + 4x_3 = 9$$
$$x_1 + x_2 - x_3 = -1$$
$$x_1, x_2, x_3 \geqslant 0.$$

Trial solution: (0, 1, 2).

7. Solve the following problems by complementary pivoting

(i) maximise $2x_1 + 2x_2 + 2x_3 - x_1^2 - x_2^2 - x_3^2$

subject to
$$x_1 + 2x_2 + 3x_3 \leqslant 3$$
$$x_1, x_2, x_3 \geqslant 0.$$

(ii) maximise $20x_1 - 10x_2 - 3x_1^2 - 3x_2^2$

subject to
$$2x_1 - x_2 \leqslant 6$$
$$-x_1 + x_2 \leqslant 10$$
$$2x_1 + 3x_2 \geqslant 8$$
$$x_1, x_2 \geqslant 0.$$

(iii) maximise $4x_1 - 4x_2 - 2x_1^2 - x_2^2 + 2x_1x_2$

subject to
$$2x_1 - x_2 = 1$$
$$x_1 - x_2 \leqslant 1$$
$$x_1, x_2 \geqslant 0.$$

8. When artificial variables are used in complementary pivoting, show that, in the $a_i$-row of the initial tableau,

(i) if $\alpha$ is the entry in the $x_j$-column, then $-\alpha$ is the entry in the $\mu_j$-row and $\lambda_i$-column,

(ii) the entry is zero in any column, if the non-basic variable is *not* $x_j$ for some $j$.

Show that pivoting as prescribed in Section 12.4 preserves properties (i) and (ii) and is therefore always possible.

# Further Reading

For the reader who wishes to pursue the study of mathematical programming further, we offer here a few suggestions for further reading.

Further study around linear programming can be pursued in at least three distinct directions. Firstly, the task of formulating real-world problems as LPs is a complex task, demanding insight and flair. Reading case studies can give a useful feeling for the difficulties involved and ways of overcoming them, and a graded set of case studies is offered by Salkin and Saha (1975). Secondly, efficient computational implementation of mathematical programming algorithms will significantly improve the practical performance of these methods. (Computer) programming and numerical–analytic aspects are covered in the collection of papers edited by Greenberg (1978). Finally, LPs can be extended by allowing non-linearities and the closest such topic to LP is convex programming. This is covered in the extensive and basic, though demanding, book of Rockafellar (1970). These last two references assume a good knowledge of finite vector spaces, and Rockafellar requires vector calculus.

Of the specialised topics covered in Chapters 6 to 12, Gal (1979) treats sensitivity analysis in great depth. The paper by Glover and Klingman (1978) includes many useful references on network problems. Zeleny (1982) discusses a multitude of approaches to multiple objective problems, but goal programming and game theory, in particular, are expounded by, respectively, Ignizio (1976) and Thomas (1984). A useful collection of papers on aspects of integer programming is edited by Hammer *et al.* (1979). Finally, Van de Panne (1975) deals extensively with quadratic programming, although the proof in Section 12.5 is a modification of that suggested by Garcia (1976).

## REFERENCES

Gal, T., 1979, *Postoptimal Analyses, Parametric Programming and Related Topics*, McGraw-Hill, New York.

Garcia, C. B., 1976, A note on a complementary variant of Lemkse's method, *Math. Prog.* **10**, 134–136.

Glover, F. and Klingman, D., 1978. Modeling and solving network problems, in Greenberg (1978).

Greenberg, H. J., (Ed.), 1978, *Design and Implementation of Optimisation Software*, Sijhoff and Noordhoff, Alphen aan den Rijn, Netherlands.

Hammer, P. L., Johnson, E. L. and Korte, B. H., (Eds.), 1979, *Discrete Optimisation* Vols. I and II, North-Holland, Amsterdam.

Ignizio, J. P., 1976, *Goal Programming and Extensions*, D. C. Heath, Lexington, M.A.

Rockafellar, R. T., 1970, *Convex Analysis*, Princeton University Press, Princeton, New Jersey.

Salkin, H. M. and Saha, J., (Eds.), 1975, *Studies in Linear Programming*, North-Holland, Amsterdam.

Thomas, L. C., 1984, *Games, Theory and Applications*, Ellis Horwood, Chichester.

Van de Panne, C., 1975, *Methods for Linear and Quadratic Programming*, North-Holland, Amsterdam.

Zeleny, M., 1982, *Multiple Criteria Decision Making*, McGraw-Hill, New York.

# Answers or Outline Solutions to Exercises

(N.B. Except where otherwise stipulated in the question, alternative optima are not given.)

**CHAPTER 1**

1. maximise $13x_1 + 12x_2$ (= profit in £1,000s)

$$\begin{aligned} \text{subject to} \quad & 2x_1 + 5x_2 \leqslant 110 \\ & 7x_1 + 2x_2 \leqslant 140 \\ & x_1 + x_2 \leqslant 30 \\ & x_1, x_2 \geqslant 0 \end{aligned}$$

where $x_1(x_2)$ is the number of days operating process I(II).
Optimal solution $(x_1, x_2) = (16, 14)$.

2. maximise $-5x_4 + 5x_5 + x_2 + x_6 - x_7$

$$\begin{aligned} \text{subject to} \quad & -x_4 + x_5 - x_2 - x_6 + x_7 + s_1 = -19 \\ & 5x_4 - 5x_5 - x_2 + 12x_6 + 12x_7 + s_2 = 10 \\ & 7x_4 - 7x_5 + x_2 - 3x_6 + 3x_7 = 5 \\ & x_4, x_5, x_2, x_6, x_7 \geqslant 0 \quad (x_1 = x_4 - x_5, x_3 = x_6 - x_7). \end{aligned}$$

3. Eliminating $x_1$ and $x_2$, for example, gives

P1: minimise $2x_3 + 2x_4$

$$\begin{aligned} \text{subject to} \quad & -x_3 - x_4 \leqslant 7 \\ & x_3 - x_4 \geqslant -1 \\ & x_3 \leqslant -1. \end{aligned}$$

Optimal vertices: $(x_3, x_4) = (-4, -3), (-1, -6)$. The midpoint of these, $(-2\frac{1}{2},$

$-4\frac{1}{2}$), is also optimal. In terms of the original problem, $(x_1, x_2, x_3, x_4) = (4\frac{1}{2}, 3\frac{1}{2}, -4, -3), (1\frac{1}{2}, 6\frac{1}{2}, -2\frac{1}{2}, -4\frac{1}{2}), (-1\frac{1}{2}, 9\frac{1}{2}, -1, -6)$ are optimal.

4.
(i) $(3, 0)$.
(ii) As $\theta$ increases, the objective function rotates clockwise, giving:

LP unbounded for $\theta < \frac{1}{2}$
$(5\frac{2}{3}, 1\frac{1}{3})$ if $\frac{1}{2} \leqslant \theta \leqslant 2$,
$(3, 0)$ if $\theta \geqslant 2$.

5. No, there can be unbounded optimal edges. For example,

$$\begin{array}{ll} \text{maximise} & x_1 - x_2 \\ \text{subject to} & x_1 - x_2 \leqslant 1 \\ & x_1, \; x_2 \geqslant 0. \end{array}$$

This has $(1 + \lambda, \lambda)$ optimal for any $\lambda \geqslant 0$, but only $(1, 0)$ is an optimal vertex.

## CHAPTER 2

1. $s_1, s_2, s_3$ basic; $x_1, x_2, x_3$ non-basic

$$\begin{aligned} x_1 + \tfrac{2}{3}x_2 + 1\tfrac{5}{6}x_2 + \tfrac{1}{6}s_2 &= 2\tfrac{1}{6} \\ x_3 - \tfrac{1}{3}x_2 - 1\tfrac{1}{6}x_2 + \tfrac{1}{6}s_2 &= \tfrac{1}{6} \\ s_3 + 1\tfrac{2}{3}x_2 + 7\tfrac{5}{6}x_2 + \tfrac{1}{6}s_2 &= 2\tfrac{1}{6}. \end{aligned}$$

2. 
$$\begin{aligned} x_1 + 1\tfrac{4}{5}x_2 - 1\tfrac{1}{5}x_4 &= -2\tfrac{1}{5} \\ x_2 + \tfrac{3}{5}x_2 - \tfrac{2}{5}x_4 &= 1\tfrac{3}{5} \end{aligned}$$

BFSs: $(-7, 2\frac{2}{3}, 0, 0), (-2\frac{1}{5}, 0, 1\frac{3}{5}, 0), (-7, 0, 0, -4), (0, -1\frac{2}{9}, 2\frac{1}{3}, 0), (0, 0, 2\frac{1}{3}, 1\frac{5}{6})$.

3. $x_1, x_2: 3$. $x_1, x_3: 5$. $x_1, x_4: -2$. $x_2, x_3: 9$. $x_2, x_4: 0$. $x_3, x_4: 6$. Determinant is zero for $x_2, x_4$ and there is no BFS with basis $\{x_2, x_4\}$.

4. For example, $\alpha'_{ir} = \alpha_{rj}/\alpha_{rk}$, $\alpha'_{ik} = -\alpha_{ik}/\alpha_{rk}$, $\alpha'_{rk} = 1/\alpha_{rk}$, so $\alpha''_{ij} = \alpha'_{ij} - \alpha'_{ik}\,\alpha'_{rj}/\alpha'_{rk} = \alpha'_{ij} + \alpha_{ik}\alpha_{rj}/\alpha_{rk} = \alpha_{ij}$.

5.
(i) Pivot in $x_5$-column, $x_3$-row; new BFS: $(\frac{1}{6}, 0, 0, 0, 2\frac{1}{6}, 1\frac{2}{3})$
$x_2$-column, $x_3$-row; new BFS: $(17\frac{1}{2}, 13, 0, 0, 0, 12\frac{1}{2})$
$x_4$-column, $x_3$-row; new BFS: $(6\frac{2}{3}, 0, 0, 2\frac{1}{6}, 0, \frac{1}{24})$

(ii) Pivot in $x_1$-column, $s_3$-row; $(s_1, s_2, s_3, x_1, x_2, x_3) = (\frac{2}{5}, 4\frac{4}{5}, 0, 1\frac{1}{5}, 0, 0)$
$x_2$-column, $s_1$-row; BFS does not change.

Can also pivot in $x_1$- or $x_3$-column and $s_2$-row. BFS does not change. $(4 + 2\lambda, 2\lambda, 6, 0, 0, \lambda)$.

6. Pivot in each column of each tableau using the PRS rule provided it gives a new tableau.

(i) $(0, 0), (3, 0), (0, 3), (2\frac{1}{2}, 1\frac{1}{2}), (1\frac{1}{3}, 2\frac{2}{3})$
(ii) $(0, 0), (0, 2), (3, 0)$.

7. $\alpha_{s0}/\alpha_{sk} = \alpha_{r0}/\alpha_{rk}$ implies $\alpha'_{s0} = 0$.

Pivoting first in row $r$ and then row $s$ of column $k$ is equivalent to a single pivot in row $s$ and column $k$. This could only lead to a primal feasible tableau if there is a tie and thus degeneracy, or $s = r$.

## CHAPTER 3

1. (i) $(4\frac{1}{2}, 3)$, (ii) unbounded, (iii) $(2\frac{3}{7}, -2\frac{4}{7}, 0)$, (iv) infeasible (v) $(2, 0, 0)$, (vi) $(0, 11\frac{1}{2})$. No alternative optima.

2. If $x_{N_j}$ is free, use $|\alpha_{0j}|$ in place of $-\alpha_{0j}$ in choosing pivot column. If $x_{B_i}$ is free, never pivot in row $i$.

3. (5, 2) is optimal. A choice of pivot row in second tableau leads to optimal tableau in two (one choice) or three pivots.

4. Maximise *and* minimise $-7x_1 + 2x_2 + 5x_3 - 4x_4 + 3x_5$ subject to the other constraints. $-9\frac{3}{5} \leqslant \alpha \leqslant 27\frac{2}{3}$.

5. (0, 0, 2, 0). The second canonical equation is $x_1 + \frac{1}{2}x_2 + \frac{1}{2}x_4 = 0$. The only non-negative solution of this is $x_1 = x_2 = x_4 = 0$. Hence $x_3 = 2$.

6. A BFS must have two of $s_1$, $s_2$, $s_3 = 0$. Only $s_1 = s_3 = 0$ is feasible. BFS: $(0, \frac{4}{5}, 0, \frac{2}{5})$. Only objective (iii) gives an objective row with consistent sign (negative). Hence *minimise* $x_1 + x_2 + x_3 - x_4$.

7. Apply the method of Exercise 6 of Chapter 2 restricted to columns with $\alpha_{0j} = 0$. Optimal BFSs: $(5, 3, 0)$, $(6\frac{4}{5}, 0, \frac{3}{5})$, $(0, 8, 0)$, $(0, 0, 4)$.

8. Unbounded set of optimal solutions.

9. Six pivots lead back to the original tableau.

10. In the notation of Section 3.6, $\alpha_{\mathrm{T}r}$ is increased by $\sum_i \mu_i \epsilon^i$ in the adjusted problem. For $\epsilon$ ($> 0$) close enough to zero, all these polynomials can be made sufficiently small not to effect the sign of any signed $\alpha_{\mathrm{T}r}$.

If $\alpha_{\mathrm{T}s}$ is increased by $\sum_i \nu_i \epsilon^i$ and $\alpha_{rk}, \alpha_{sk} > 0$, row $r$ is preferred to row $s$ if $\alpha_{\mathrm{T}r}/\alpha_{rk} < \alpha_{\mathrm{T}s}/\alpha_{sk}$, or these ratios are equal and $\mu_1/\alpha_{rk} < \nu_1/\alpha_{sk}$, or both these ratios are equal and $\mu_2/\alpha_{rk} < \nu_2/\alpha_{sk}$ or . . . . The refined PRS rule selects row $r$ which is preferred to all others with $\alpha_{rk} > 0$. This must give a unique choice or degeneracy would occur. The $\mu_1, \ldots, \mu_m$ are carried in extra columns and these columns are transformed as usual.

Optimal solution: $(\frac{1}{2}, 0, \frac{1}{2}, 0)$.

## CHAPTER 4

1. $(0, 0, \frac{3}{7}, \frac{2}{7}, 0, 0, 0, \frac{2}{7}, 0)$.

2. $(1\frac{1}{2}, 2, 0, 0, 0, 0, 2\frac{1}{2}, 0, 0, 0, 0, 0)$.

3. Either $\alpha_{r0} \neq 0$, in which case there is no solution (with $x_{B_r} = 0$ and $x_{N_j} = 0$ for $j \neq k$), or $\alpha_{r0} = 0$, in which case $x_{N_k} = \lambda$, $x_{B_i} = \alpha_{i0} - \lambda\alpha_{ik}$ for all $i$, is a solution for any $\lambda$. The existence of the presumed basis would imply a unique (basic) solution. Hence, each column must have a non-zero entry in a suitable row in each remaining column in the reinversion procedure.

4. See Exercise 1.

5. (ii) requires one less iteration than (i). Optimal solution: (0, 1.2).

**CHAPTER 5**

1. (i) optimal, (ii) not optimal, (iii) not optimal.

2. Optimal dual solution: $(y_1, y_2) = (1, 2)$. Optimal primal solution: $(0, 0, \frac{1}{4}, 2\frac{1}{4})$.

3. $x_3, s_1, s_3 > 0$ imply $t_3 = y_1 = y_3 = 0$ in dual constraints, giving P7. P7 has no feasible solution, so (0, 0, 2) is not optimal in P6.

4. (i) $(2, \frac{1}{4}, \frac{1}{4})$ would imply $y_3 = -\frac{1}{7}$ in dual constraints. Hence, not optimal. (ii) If $\beta > 3$, CS conditions imply $3y_1 = 1 + \alpha$, $3y_3 = 2 - \alpha$, $3t_3 = 4\alpha - 8$. So $y_1, y_3, t_3 \geqslant 0$ mean $\alpha = 2$.

5. (i) $(0, 11\frac{1}{2})$, (ii) infeasible.

6. $\alpha_{rk} < 0$ ensures $\alpha'_{0k} \geqslant 0$ etc.

7. $\beta_{rj} = \sum_i \pi_i a_{ik}$ if $X_j$ is $x_k$ and $= \pi_i$ if $X_j$ is $s_i$, where $\pi_1, \ldots, \pi_m$ are obtained by backwards transformation starting with $e_r = 1, e_i = 0$ for $i \neq r$.

8. Optimality says $\beta_{0j} \geqslant 0$, i.e. $\sum_i \pi_i a_{ik} \geqslant c_k$, all $k$, and $\pi_i \geqslant 0$, all $i$. Also, $\beta_{00} = \sum_i \pi_i b_i$ says P and D objective function values are equal.

9. First part: obvious. Deduction: if P has a non-empty bounded feasible region, it can be made infeasible by adjusting the RHSs of the constraints. D is feasible (why?) and the adjustment, affecting only the objective function of D, makes D unbounded. The result follows.

**CHAPTER 6**

1. (i) (1, 1, 1); (ii) $(0, 2, \frac{1}{2})$; (iii) use two-phase method to restore optimality; (1, 1, 1); (iv) $(0, 2\frac{3}{7}, \frac{5}{7})$; (v) $(0, 2\frac{1}{4}, \frac{5}{8})$; (vi) infeasible; (vii) add a constraint to the dual problem; $(0, 3\frac{7}{8}, \frac{3}{4}, 1\frac{3}{8})$; (viii) drop $s_1$-row; $(0, 2\frac{1}{2}, \frac{3}{4})$; (ix) make $s_2$ basic, drop $s_2$-row, restore optimality; (0, 4, 0) or $(\frac{2}{3}, 3\frac{1}{3}, 0)$; (x) $x_2 = 2\frac{1}{2}, x_3 = \frac{3}{4}$; (xi) add $x_2 \leqslant 0; x_1 = \frac{3}{4}, x_3 = \frac{1}{4}$.

2. $c_1 \leqslant 1\frac{1}{2}, c_2 \geqslant 1\frac{2}{3}, -2 \leqslant c_3 \leqslant 2$.

3.

(i) $(0, 2\frac{1}{2}, \frac{3}{4})$ for $\theta \leqslant \frac{1}{2}$,

$(1, 1, 1)$ for $\theta \geqslant \frac{1}{2}$.

(ii) $(\frac{1}{2}, 0, 0)$ for $\theta \leqslant -1$,

$(\frac{3}{4}, 0, \frac{1}{4})$ for $-1 \leqslant \theta \leqslant -\frac{5}{7}$,

$(1, 1, 1)$ for $-\frac{5}{7} \leqslant \theta \leqslant -\frac{2}{5}$,

$(0, 2\frac{1}{2}, \frac{3}{4})$ for $\theta \geqslant -\frac{2}{5}$.

(iii) $(0, 2\frac{1}{2}, \frac{3}{4} - \frac{1}{2}\theta)$ for $-1\frac{5}{6} \leqslant \theta \leqslant 1\frac{1}{2}$,

$(-3 + 2\theta, 7 - 3\theta, 0)$ for $1\frac{1}{2} \leqslant \theta \leqslant 1\frac{5}{6}$,

$(\frac{2}{3}, 3\frac{1}{3} - \theta, 0)$ for $1\frac{5}{6} \leqslant \theta \leqslant 3\frac{1}{3}$,

$(0, -10 + 3\theta, 0)$ for $3\frac{1}{3} \leqslant \theta \leqslant 4$,

infeasible for $\theta < -1\frac{5}{6}$ or $\theta > 4$.

4.

(i) $(0, 0, 0, 2, 0)$ for $\theta \leqslant -\frac{2}{11}$,

$(1\frac{1}{3}, 0, \frac{2}{3}, 0, 0)$ for $-\frac{2}{11} \leqslant \theta \leqslant 14$,

$(2, 2, 0, 0, 0)$ for $\theta \geqslant 14$.

(ii) $(0, -4 - 1\frac{1}{2}\phi, 2 + \frac{1}{8}\phi, 0, 0)$ for $-16 \leqslant \phi \leqslant -2\frac{2}{3}$,

$(1\frac{1}{3} + \frac{1}{2}\phi, 0, \frac{2}{3} - \frac{3}{8}\phi, 0, 0)$ for $-2\frac{2}{3} \leqslant \phi \leqslant 1\frac{7}{9}$,

infeasible for $\phi < -16$ or $\phi > 1\frac{7}{9}$.

5. $(x_1, x_2, x_3) = (4 + \theta, 0, 0)$ for $-4 \leqslant \theta \leqslant -2$,

$= (0, 2 + \frac{1}{2}\theta, 0)$ for $-2 \leqslant \theta \leqslant 1\frac{1}{5}$,

$= (0, 5 - 2\theta, -6 + 5\theta)$ for $1\frac{1}{5} \leqslant \theta \leqslant 2$,

$= (0, 0, 4 + \theta)$ for $2 \leqslant \theta \leqslant 2\frac{1}{2}$,

infeasible for $\theta < -4$ or $\theta > 2\frac{1}{2}$.

6.

(i) $x_1 = \theta = 4, x_2 = 0$ is feasible

$(1\frac{1}{7} + \frac{5}{7}\theta, 1\frac{5}{7} - \frac{3}{7}\theta)$ for $\frac{1}{2} \leqslant \theta \leqslant 4$,

$(8 - \theta, 0)$ for $4 \leqslant \theta \leqslant 5$,

infeasible for $\theta < \frac{1}{2}, \theta > 5$. (N.B. includes $\theta = 0$.)

(ii) The dual problem is feasible for $\theta = \frac{4}{7}$ ($y_1 = 0, y_2 = \frac{3}{7}$), so start with this.

(4, 3) for $\frac{4}{7} \leqslant \theta \leqslant 1$,

(1, 0) for $\theta \geqslant 1$,

unbounded for $\theta < \frac{4}{7}$.

(iii) The dual problem is feasible for no $\theta$. So LP is unbounded for all $\theta$.

7. $(-\theta, 0)$ for $-3 \leqslant \theta \leqslant 0$,

$(-1\frac{1}{2} - 1\frac{1}{2}\theta, 0)$ for $\theta \leqslant -3$,

infeasible for $\theta > 0$.

## CHAPTER 7

1. (i) (3, 8, 3). (ii) (1, 4, −1, 0, 2).

2. (a) $(2\frac{1}{2}, 8, 2)$. (b) (3, 1, 0). (c) (2, 4, 6). (d) Infeasible.

3. Entries are the same except in the pivot column, where the sign difference between the $x_J$- and $\sigma_J$-rows forces the sign change. With the PFRS method we can keep the same $\eta$-list but remember to multiply the column corresponding to $X_j$ by $-1$ (including the price) if $X_j$ is $\sigma_J$.

## CHAPTER 8

1. (i) $x_{23} = 1$, $x_{43} = 1$, $x_{53} = 2$, $x_{61} = 8$, $x_{63} = 1$. (ii) $x_{21} = 1$, $x_{23} = 0$, $x_{41} = 1$, $x_{53} = 5, x_{61} = 9$.

2. The solution satisfies $\sum_{j \neq i} x_{ji} - \sum_{k \neq i} x_{ik} = b_i$ for $i$ in $T_A$, where the sums are over $T_A$. Summing over $i$ gives the result. With the suggested changes, no partial sum can be zero, if $m\epsilon < 1$ (why?).

3. Suppose $r = G(J) - G(I) > 0$. Let $J_0 = J$, $J_t = P(J_{t-1})$ for $t \geqslant 1$. Compare $J_t$ and $I$, $J_{t+1}$ and $I_1 = P(I)$, $J_{t+2}$ and $I_2 = P(I_1)$, . . . , until these coincide at $K$, say. The path is $J$, $J_1$, $J_2$, . . . , $K$, . . . . , $I_2$, $I_1$, $I$. Can also cope with $G(J) \leqslant G(I)$.

4.

(i) $x_{14} = 5, x_{21} = 1, x_{23} = 6, x_{24} = 2, x_{31} = 4, x_{32} = 2, x_{35} = 3$, other $x_{ij} = 0$.

(ii) $x_{32} = 5$, others the same as (i).

(iii) $x_{11} = 2, x_{14} = 7, x_{21} = 3, x_{23} = 6, x_{32} = 2, x_{35} = 3$, other $x_{ij} = 0$.

5. Using same tree as 4(i), find $x_{14} = 6, x_{21} = 2, x_{24} = 1, x_{31} = 3, x_{35} = 4$, others as 4(i). Original cost $= 83$. New cost $= 80$. So more is transported at less cost.

6. The initial solution is optimal.

7. Set $c_{ij} = A + h\,(j-1)$ if $j < i$,
$= 0$ if $j = i$,
$= K$ (a large number) if $j < i$.

Must have $\sum_{1}^{n} a_i \geqslant \sum_{1}^{n} b_i$ for $n = 1, \ldots, N$.

8. $x_{14} = x_{22} = x_{33} = x_{41} = x_{55} = 1$.

9. $x_{21} = 1$, $x_{41} = 3$, $x_{51} = x_{54} = 1$, $x_{61} = 3$, $x_{63} = 5$, $x_{64} = 1$, other $x_{ij} = 0$.

10. If $U_{ij} < \min\{\alpha_i, \beta_j\}$ in the initialisation procedure for a TP, set $x_{ij} = U_{ij}$ ($\sigma_{ij}$ non-basic), subtract $U_{ij}$ from $\alpha_i$ and $\beta_j$, and consider $x_{ij}$ no longer available, unless $x_{ij}$ is the last remaining entry in column to be filled (in which case proceed as normal). Starting solution: $x_{12} = x_{25} = x_{33} = 3$, $x_{12} = x_{21} = 2$, $x_{24} = 1$, $x_{14} = 6$ ($x_{33} = 3$ is non-basic). Optimal solution: $x_{11} = x_{32} = 2$, $x_{14} = x_{21} = x_{23} = x_{24} = x_{33} = x_{35} = 3$, $x_{34} = 1$, other $x_{ij} = 0$.

## CHAPTER 9

1. (i) Not efficient, dominated by (3, 3). (ii) Efficient. (iii) Efficient.

2. (i), (iii) No efficient solutions. (ii) $(2\frac{1}{4}, \frac{1}{2})$. (iv) $(0, 0)$.

3. $(y_1, \ldots, y_m)$ feasible in the dual problem of LP$(1 + v_1, \ldots, 1 + v_p)$ means $(v_1, \ldots, v_p, y_1, \ldots, y_m)$ feasible in the dual problem of LP$^*$ $(\hat{x}_1, \ldots, \hat{x}_n)$. The result follows.

4. Exercise 2(ii): only $(2\frac{1}{4}, \frac{1}{2})$.
Exercise 2(iv): $(\phi, 0)$ for $0 \leqslant \phi \leqslant \frac{2}{3}$,
$(\frac{2}{3} + \frac{2}{3}\phi, \phi)$ for $0 \leqslant \phi \leqslant 2$.

5. $(4, 2\frac{1}{2})$ is the only efficient solution, but $(5\frac{1}{4}, 0)$ is weakly efficient.

6. Any solution dominating $(\hat{x}_1, \ldots, \hat{x}_n)$ would also be optimal in LP$(w_1, \ldots, w_p)$.

7.

| | $s_2$ | $x_2$ | $x_1$ | |
|---|---|---|---|---|
| $x_3$ | $-\frac{1}{8}$ | $\frac{1}{4}$ | $\frac{1}{8}$ | $\frac{3}{4}$ |
| $x_1$ | $\frac{5}{8}$ | $\frac{1}{4}$ | $-\frac{1}{8}$ | $\frac{3}{4}$ |

This tableau is optimal.

8. $(2, 0, 0), (0, 0, 2)$.

## CHAPTER 10

1. (i) $(\frac{15}{11}, \frac{35}{22})$. (ii) $(\frac{1}{4}, 10\frac{1}{2})$.

2.

| | |
|---|---|
| $(0, 0)$ | for $\theta \leqslant -10$ |
| $(0, 10+\theta)$ | for $-10 \leqslant \theta \leqslant -9\frac{1}{2}$, |
| $(9\frac{1}{2}+\theta, \frac{1}{2})$ | for $-9\frac{1}{2} \leqslant \theta \leqslant -5\frac{5}{6}$, |
| $(-8-2\theta, 18+3\theta)$ | for $-5\frac{5}{6} \leqslant \theta \leqslant -5\frac{5}{7}$, |
| $(3\frac{3}{7}, \frac{6}{7})$ | for $\theta \geqslant -\frac{5}{7}$. |

3. (i) and (ii) as in Sections 10.1 and 10.3.
(iii) maximise $2\phi + 5d_1 + 5e_1$

subject to

$$\begin{aligned}
d_2 + e_2 + 2d_3 + 2e_3 &\leqslant 12 \\
4x_1 + 3x_2 &\geqslant 7 \\
3x_1 + 6x_2 &\geqslant 7 \\
x_1 + 3x_2 - d_1 + e_1 &= 7 \\
x_1 + 9x_2 - d_2 + e_2 &= 0 \\
4x_1 + 3x_2 - d_3 + e_3 &= 0 \\
x_1, x_2, d_j, e_j \geqslant 0 \quad &\text{for } j = 1, 2, 3.
\end{aligned}$$

4. $(p_1, p_2, p_3) = (\frac{1}{2}, \frac{1}{2}, 0), (q_1, q_2, q_3) = (\frac{1}{2}, 0, \frac{1}{2})$, value $= 0$.
If I plays $(\frac{1}{2}, \frac{1}{2}, 0, 0)$ in extended game he wins at least zero; if II plays $(\frac{1}{2}, 0, \frac{1}{2}, 0)$ he loses at most zero. Hence these are optimal.

5. Number the strategies 1: show 1, guess 1; 2: show 1, guess 2; 3: show 2, guess 1; 4: show 2, guess 2. Then $a_{12} = 2, a_{13} = 3, a_{24} = 3, a_{34} = -4$, other $a_{ij} = 0$, if $i < j$. Verify $\sum_i a_{ij} x_i \geqslant 0$, with $x_2 = \frac{7}{12}, x_3 = \frac{5}{12}, x_1 = x_4 = 0$.

6. Value $= \max\{-\frac{1}{9}, -\frac{1}{9}, -\frac{1}{9}, -3\frac{4}{9}\} = -\frac{1}{9}$. Add 9 to all entries.
The CS conditions can be used to show $(\frac{1}{3}, \frac{2}{9}, 0, \frac{4}{9})$ is optimal in LPII. maximin for I: $(\frac{1}{2}, \frac{4}{9}, \frac{1}{18}, 0)$.

## CHAPTER 11

1. (1, 1). No rounded solution is feasible.

2. (i) (1, 4). (ii) (0. 0, 3).

3. Neither problem has alternative optima.

4. (4, 0, 4).

5. The solution $x_{N_k} = \lambda$, $x_{B_i} = \alpha_{i0} - \alpha_{ik}\lambda$, for $i \geqslant 1$, has integer values if $\lambda$ is any multiple of the denominators of all the $\alpha_{ik}$.

6. (1, 0, 2, 1) has $z = 29$. $z^* = 30$.

7. $(x_1, x_2) = (\frac{2}{3}, 2)$.

8. $1 \to 4 \to 3 \to 2 \to 5 \to 6 \to 1$.

## CHAPTER 12

1. (i) (3, 0). (ii) $(-\frac{1}{3}, 6\frac{1}{3})$. (iii) $(1, 3\frac{3}{5}, \frac{2}{5})$.

2. The feasible region is a closed, bounded set and the objective function is continuous.

3. (i) and (iii) are PSD.

4. Set out $q_{jk}$ as a tableau. Pivot in row 1, column 1 provided the pivot is positive. Drop row 1 and column 1 and pivot in row 2, column 2 etc. Stop if a negative pivot or a zero pivot in a non-zero row is encountered (not PSD).

5. Using the method of Section 12.3 the objective function of any QP can be written as

$$\sum_j d_j y_j - \sum \delta_j y_j^2 = -\sum_j \delta_j (y_j - \tfrac{1}{2} d_j \delta_j^{-1})^2 + \tfrac{1}{4} \sum_j d_j \delta_j^{-1},$$

where the PD property guarantees $\delta_j > 0$. Clearly QP is bounded and therefore has a (unique) optimal solution by Exercise 2.

6. The KT conditions are satisfied in (i) and (ii) and they are concave. The KT conditions are not satisfied for (iii).

7. (i) $(\frac{11}{14}, \frac{8}{14}, \frac{5}{14})$. (ii) $(3\frac{1}{4}, \frac{1}{2})$. (iii) $(\frac{1}{2}, 0)$.

8. (i) and (ii) from KT conditions. The first pivot in each pair of pivots only affects $x_j$-columns and the second only the $\mu_j$-rows. So (ii) is preserved. Then (i) follows from the transformation rules. We cannot encounter a zero pivot because $\alpha \neq 0$ implies $-\alpha \neq 0$.

# Index